Biophilia and Anti-Immortalism in Children's Literature

Children's literature has traditionally been analyzed through humanist frameworks like psychoanalysis and gender theory, which, while legitimizing the genre's study, fail to address its focus on animals, rural settings, and nature. This project takes a non-humanist, sociobiological approach, arguing that children's literature has, for over a century, fostered what E. O. Wilson calls "biophilia"—an evolution-based affinity for living things expressed through innate impulses and predispositions. Focusing on chapter books with immortality as a theme, such as Frances Hodgson Burnett's *The Secret Garden*, Natalie Babbitt's *Tuck Everlasting*, and Ursula Le Guin's *The Other Wind*, the study explores how children's literature promotes a syncretic biotheism, revering life's natural cycles while rejecting immortalist ideologies like Christian post-death immortality and secular life extension. These books depict human death as reintegration into nature, offering young readers a unique "ecological literacy." This literacy cultivates values that conflict with industrialism, urbanization, and Christianity, creating "internal anomie" as readers transition into adulthood. By examining how children's literature shapes ecological awareness and challenges dominant cultural mores, the project highlights the genre's role in fostering a deeper connection to the natural world and its cycles, often overlooked in traditional critical analyses.

Pamela Swanigan is a writer, editor, and academic coach in Vancouver, British Columbia.

Routledge Studies in World Literatures and the Environment
Series Editors: Scott Slovic and Swarnalatha Rangarajan

Reading the Environment
Olga Tokarczuk's Fiction
Dominika Oramus

Ecomaterial Ambivalence and Environmental Crises in South and Southeast Asian Literary, Philosophical and Cultural Narratives
Chitra Sankaran

The Environment in Sustainable American Studies
Powered by Nature
Edited by Frank Mehring

The Anthropocene Sublime in Literature
Immanence and Entanglement from Romanticism to Digitality
Thomas Storey

The Environmental Graphic Novel
How Green Is the Panel
Pramod K. Nayar

John Clare's Modern Eco-Poetic Legacy in English
Sam Hickford

Environmental Transformations in Victorian, Edwardian and Modernist Essays
Edited by Dominika Buchowska-Greaves and Christine Reynier

Biophilia and Anti-Immortalism in Children's Literature
Born to Die
Pamela Swanigan

For more information about this series, please visit: www.routledge.com/Routledge-Studies-in-World-Literatures-and-the-Environment/book-series/ASHER4038

Biophilia and Anti-Immortalism in Children's Literature

Born to Die

Pamela Swanigan

Routledge
Taylor & Francis Group
NEW YORK AND LONDON

First published 2027
by Routledge
605 Third Avenue, New York, NY 10158

and by Routledge
4 Park Square, Milton Park, Abingdon, Oxon, OX14 4RN

Routledge is an imprint of the Taylor & Francis Group, an informa business

ISBN: 978-1-041-32728-8 (hbk)
ISBN: 978-1-041-32730-1 (pbk)
ISBN: 978-1-003-78421-0 (ebk)

DOI: 10.4324/9781003784210

Typeset in Sabon
by Deanta Global Publishing Services, Chennai, India

For Kate, who mysteriously and magically knows the right thing to say at all times. "Save it for the book" was exactly what I needed to hear when I was inclined to revise dissertation chapters into the next incarnation. I'm sure she foresaw, as came to pass, that by the time the book was upon me, I would have forgotten every change I meant to make and would meander through the process in a state of amnesiac contentment.

For Sam Pickering, who early on commanded me, with an emphatic slap of his desk, "Pamela, *lower your standards!*" His faith that I could do so and still produce a work of quality sustained me through many hours of self-doubt.

And of course for Matt. We have been through the wars together, as the saying goes, and this project was one of the longest battles, undertaken at a time of poverty, isolation, and fear. Thank you for being my unwavering champion and beloved companion, and for being with me yet.

Contents

Introduction

Strange non-bedfellows: How children's literature and ecocriticism managed not to get together

In his celebrated 1991 ecocritical reclamation of William Wordsworth, *Romantic Ecology*, British scholar Jonathan Bate casts a wry eye over the work of his fellow Romantics scholars and finds it due for some gentle upbraiding. "[S]ome of the most eminent literary critics of our time," he notes in his introduction,

> have believed that Wordsworth was not a nature poet, or that there is no such thing as nature, or that if there is such a thing and Wordsworth was interested in it then that interest was very suspect on political grounds. The common reader's view of Wordsworth derives from the Victorian way of reading him, John Stuart Mill's way, John Ruskin's, Leslie Stephen's; this book will argue that, unfashionable as that way now is in literary circles, it might just be the most useful way of approaching Wordsworth in the 1990s and the early twenty-first century.[1]

We might, on a dull day, perform the experiment of scribbling substitute terms about children's literature into Bate's observation. The result might read something like this:

> Some of the most eminent children's literature critics of our time have believed that children's literature is not a nature-oriented genre, or that there is no such thing as nature, or that if there is such a thing and children's literature is centrally concerned with it, then that concern is very suspect on psychoanalytical grounds.

Our new statement would illuminate as unfortunate a truth as does Bate's original. Children's literature surely constitutes by far the largest repository of nature-focused writing to be found in Western culture. The great majority of books in the canon and on the bestsellers' backlist are explicitly about nature, animals, the fascinations of countryside living, the wonders of the frontier, and so on. These include *Charlotte's Web*, *The Secret*

DOI: 10.4324/9781003784210-1

Garden, *The One and Only Ivan*, the Little House series, *Heidi*, *Bambi*, *Rascal*, *King of the Wind*, *Island of the Blue Dolphins*, *Black Beauty*, *Trumpet of the Swan*, *Sounder*, *Goodnight, Mr. Tom*, *The Lord of the Rings*, *The Incredible Journey*, *The Jungle Book*, The Borrowers series, the Prydain Chronicles, the Middle Earth books, *Impossible Creatures*...The list goes on almost endlessly. In fact, it would take less time to name children's books that aren't animal- or nature-focused than ones that are.

Beyond the wealth of animal stories and more general nature-focused works, even many canonical children's texts that are not so obviously "nature writing" can be considered such. For instance, Elizabeth Enright's Melendy quartet, most of the books in Cynthia Voigt's Dicey series, and the Hunger Games trilogy qualify as "environmental texts" under all four of ecocritic Laurence Buell's core criteria: they depict human history as being involved in natural history, nonhuman interests as being of value, humans as being accountable in their interactions with the environment, and nature "as a process...rather than as a constant or a given."[2]

Yet, as any survey of children's literature criticism will show, little scholarship exists that focuses on this aspect of the genre, and that which does tends to treat nature as "suspect on psychoanalytic grounds." Kenneth Kidd notes that "the serious study of children's literature began with Freud"[3] and that over the years children's literature studies and psychoanalysis have been "mutually enabling,"[4] or even "mutually constitutive."[5] Thus born into the era of Ricoeurian "hermeneutics of suspicion,"[6] the critical corpus stood little chance of offering many of those happy correspondences when, in Bate's words, "the critic's purposes are also the writer's, and ... there can be a communion between living reader and dead writer which may bring with it a particular enjoyment and a perception about endurance."[7]

Here, for instance, is Karen Coats, performing an extended examination of one of the bestselling children's chapter books of all time,[8] E. B. White's *Charlotte's Web*. "As a pig, [Wilbur] is not especially important to [Fern]... He is simply a symbol with whom she narcissistically identifies." Thus, when her father sets out to kill the runt piglet, "acting... as the primordial Freudian father, exercising complete control over who may or may not acquire the phallus in his wee tribe," Fern is motivated not by any caring for the animal but only for her own sexual power when she "loses control and grabs her father's, um, ax."[9]

While this analysis may satisfy those who subscribe to Freudian theory, it doesn't accord with White's concerns about *Charlotte*. These ran to sending illustrator Garth William books on spiders so that he could accurately render the *Aranea cavatica* in White's Maine barn who served as the real-life Charlotte[10] and to considering the critiques of his step-granddaughter Caroline, who said, "Charlotte should have told [Wilbur] not to eat, then he wouldn't have been killed because he would have been too thin."[11] When White was asked to explain his own Fern-like rescue of a pig some years before he finished *Charlotte's Web*, as detailed in his 1948 *Atlantic* magazine essay "Death of a Pig," he told a critic, "[It] is a straight narrative...I was pursued by the shadow of the irony (or perhaps idiocy) of a man's desire to save the life of a creature he had every intention of murdering."[12] If analytical approaches such as Coats' fail to line up with the authors' own conceptions of their works, it is safe to expect that they line up even less with the "common reader's" experience of the text.

But then what *is* the view of the "common" child reader, free of the hermeneutics of suspicion or, most likely, hermeneutics of any kind? What cumulative worldview does she, he, or they get from reading book after book about dogs and cats, farm life, the open prairie; about animals that talk and trees that walk; about what it might feel like to be an orphaned deer, an abused horse, a free-living mole, an encaged gorilla? Does this reading "derive," generally speaking, from any one direction or source, as Bate claims for readers of Wordsworth? And, as a related question, how does this ingestion of nature-rich books affect what Sidney I. Dobrin terms "ecological literacy"?[13] That is, what do children's texts teach the child reader about the natural world, whether by omission or commission?

Simply put, I believe that children take textual depictions of affinity for nature at face value. I believe that their attraction to nature in books is in a literal sense natural, and that their way of reading both stems from and affirms their identity as members of the animal kingdom and the living world. If we admit this premise, we can view the preponderance of texts for children as lodging Western children's literature as a whole within a biological rather than a humanist or Judeo-Christian paradigm.

In fact, once we clear away the distortions of psychoanalytic and humanist lenses, it is plain that the genre works from a premise that we currently would term "sociobiological." Sociobiology is defined as regarding humans as an animal species that evolved within, belongs to, and depends upon the natural world. A large number of children's books, from bath books and picture books to chapter books and (to a lesser extent) young adult books, promote the sociobiological proposition that, as David J. Buller writes, "human culture as a whole is not opposed to biology, but is part of it," and "the practices that constitute human cultures differ only in degree of complexity, not in kind, from the web-spinning habits of spiders."[14] By almost invariably presenting children's relationships with animals and other aspects of nature as cherished and meaningful, these books echo the belief of sociobiology founder E. O. Wilson that "the phylogenetic continuity of life with humanity...does not diminish humanity—it raises the status of nonhuman creatures."[15]

It is further evident that, from these biological and sociobiological foundations, children's literature helps to amplify within the child reader the hardwired affinity for nature that Wilson has termed "biophilia." The biophilia hypothesis proposes, in a nutshell, that "We are in the fullest sense a biological species, and will find little ultimate meaning apart from [i.e., separate from] the remainder of life,"[16] The theory posits that we humans have certain predispositions based on human evolution, including "an innate tendency to focus on life and lifelike processes"[17] and the ability to "learn particular things about certain kinds of plants and animals" with "quickness and decisiveness."[18] Wilson observes, "[T]he brain appears to have kept its old capacities, its channeled quickness. We stay alert and alive in the vanished forests of the world."[19]

Given that a large number of children's books depict people as having not only the above biophilic traits but many others as well, we could suspect the genre of fostering within the Western child reader a strongly biophilic outlook. I believe that this is exactly what it does. In fact, I believe that in consistently depicting children as the young of an animal species and therefore a part of the natural world, and in containing storylines that show that "animals may be social others to us with whom we can form relationships,"[20] modern Western children's literature creates a separate and subversive culture for child readers. This culture presents, firstly, an ethical set sharply at odds with the human-exceptionalist creeds of both humanism and techno-industrialism; and, secondly, alternatives to Judeo-Christianity, in the form of biotheism, syncretism, natural immanence, and other pre- or non-Christian modes of spirituality.

When we say that children's literature promotes in the child reader a biological and biospiritual understanding of the world, we've said a lot. In fact, we've assigned the genre a radical, countercultural status. Val Plumwood, in *Feminism and the Mastery of Nature*, briefly puzzles over why "the apparent truism [of humans being part of nature] should find so much resistance and should need to be stressed so much."[21] But she answers herself almost immediately: "[T]he reason why this message of continuity and dependency is so revolutionary in the context of the modern world is that the dominant strands of western culture have for so long denied it, and given us a model of human identity as only minimally and accidentally connected with the earth."[22]

The "model" she refers to is, of course, the Judeo-Christian one. As Lynn White, Jr., argues in his important essay "The Historical Roots of Our Ecological Crisis," Judeo-Christianity is "the most anthropocentric religion the world has seen,"[23] and in that anthropocentricity, it created the preconditions for industrial capitalism: "By destroying pagan animism... [Christianity] made it possible to exploit nature in a mood of indifference to the feelings of natural objects."[24]

The mention of human non-animality brings us to the topic of immortality. More than any other concept, the idea that humans can live forever symbolizes not only a denial of our animal identity but a disavowal of it. We can view it as the antithesis of Wilson's biophilia hypothesis, since it proposes that humans are totally outside the natural cycle of life and separate from all other living beings by virtue of our immunity to death. When eternal life is cast as a desirable condition, human immortality places us both apart from and above the rest of the living world. Hence, the concept or promise of human immortality has long served as a chief postulate of human exceptionalism for Western culture's two dominant doctrines: Christianity and its secular outgrowth, the rigidly empiricist mode of science and technology aptly referred to as "scientism."

Immortality can be said to come in two types. The first is the post-death Christian immortality of eternal life for those who reach Heaven. The second is *pre*-death, incomplete-death, or immediately post-death immortality through scientific or technological means. J. Robert Adams, performing what he terms a "sensible" search for life after death, encapsulates the second approach thusly: "Can the Grim Reaper be held at bay? The answer is, yes. How? By halting the aging process, use of stem cells, nanobots, freezing, gene splicing ... nerve impulses, electromagnetism, [and] quantum mechanics."[25] Perhaps because the afterlife form of immortality is part of a larger belief system, it is the pre-death version, where immortality itself is the belief system and the aspiration, to which the term "immortalism" is generally applied. Likewise, the term "immortalist" usually refers to a person who holds the goal of living indefinitely to be both desirable and achievable.

Children's literature rejects both forms of immortality as it would a foreign body. However, it shows particular alarm at the immortalism form, and particular repugnance at immortalists *per se*. The somewhat greater acceptance of post-death immortality lines up with the genre's sociobiological and biophilic inclinations in two ways. First, post-death immortality accepts the biological reality of humanity at least to the extent that the physical body dies. Second, the immortality part of the story happens offscreen, as it were, and can be configured by the author's imagination. More often, as in the Summer Country of Lloyd Alexander's Prydain Chronicles or the Valinor (Undying Lands) of Tolkien's *Lord of the Rings*, the post-death realm is lightly sketched and otherwise left to the reader's own imagination. (It is worth noting that such descriptions tend heavily toward Edenic natural settings, rather than being urban or human-focused.) The concept of the Christian afterlife therefore presents a weaker threat to biophilic impulses.

Pre-death immortality has a harder row to hoe, because its proponents offer it as the literal ability for a person to avoid dying. The proposition offends biological sensibilities not only in arrogating humans but in ignoring an overwhelming body of evidence—for, as far as anyone can tell, every person in human history has died. As Simon Tugwell puts it,

> It is self-evident that humans are liable or subject to death and are therefore not plausible candidates for immortality, unless we are willing to envisage a being which is simultaneously mortal and immortal ... In light of the plain fact that humans die, and are therefore mortal, it is clear that some kind of conceptual jiggery-pokery must be involved in the contention that they are immortal.[26]

Children's books reject non-dying people so forcefully and consistently as to suggest a deep emotional revulsion and moral horror at the thought. The child reader experiences the nightmarish Cauldron-Born of the Prydain Chronicles, the monstrous Sauron in *The Lord of the Rings*, the sinister puppet-man of *Tuck Everlasting*, the *gebbeths* of the Earthsea cycle, and so on.

Yet this is by no means to say that immortality in every form is absent from children's literature. On the contrary, many books present spiritual aspects of being a biological human—responses of awe, reverence, mystery, and ineffable communion with the rest of the natural world—that take those books beyond the function of affirming the child's hardwired affinity for life and into that of promoting biospirituality. In this regard, we might consider children's literature within Western culture as a crucial keeper of what Native American advocate John Collier calls "the long hope": the hope that overlooked pockets in the Westernized world will succeed in resisting biophobic, industrialist, and post-industrialist values, thereby keeping alive an ember of the "ancient, lost reverence and passion for the earth and its web of life."[27]

The nature of the project

In this book I examine British and American children's novels that approach human immortality from an ecocentric or biological perspective. Broadly speaking, they exhibit this perspective in two ways. Firstly, they condemn both religious and secular philosophies that promote a yearning and striving for eternal or unnaturally extended life. Secondly, they offer alternative paradigms and attitudes, ones that recast death as a way to participate in the natural cycle and reunite with the rest of the living world.

Some of the books, such as Frances Hodgson Burnett's *The Secret Garden* and Diana Wynne Jones' *Drowned Ammet*, do the first by doing the second. That is to say, they implicitly condemn technological or religious immortalism by presenting alternative natural or biospiritual depictions of immortality. Others, such as Ursula K. Le Guin's *The Farthest Shore* and Natalie

Babbitt's *Tuck Everlasting*, indict immortalism explicitly and sternly, and enjoin the child reader to accept death as a natural inevitability.

The children's novels in this book both challenge the tenets of immortalist thinking and offer ways in which we might conceive of humans as being simultaneously mortal and immortal. Some take on the anti-nature implications of the eternal afterlife promised by the Christian belief system. Others reconfigure and reconceive immortality in ways that propose an ecologically integrated rather than segregated humanity.

In the course of examining these perspectives on immortality, I also analyze how they represent the books' larger "nature-endorsing" positions[28] and their biological and biospiritual view of humans—in other words, their depiction of humans as animals, with much of the higher meaning and deeper gratification of life deriving from interactions with the biotic and abiotic natural matrix. I explore two key possibilities that arise from this premise. The first is that this strain of children's literature acts as a smuggler of pre-Industrial and non–Judeo-Christian perceptions and beliefs down through the generations in a "safe" vehicle. As Kimberley Reynolds notes, the genre is "a curious and paradoxical cultural space: a space that is simultaneously highly regulated and overlooked, orthodox and radical, didactic and subversive."[29]

The second possibility, following from the first, is that these ecocentric children's texts help to create in modern Western readers a high degree of internal anomie—a sharp conflict between values and allegiances acquired in childhood and those socially imposed in adulthood. This idea is hardly new. In fact, it works us right back around to Rousseau's distinction between the state of the human character before and after losing touch with "Nature within us,"[30] and echoes the Wordsworthian creed that "Heaven lies about us in our infancy!" but "Shades of the prison-house begin to close upon the growing boy."[31] It also accords with Karl Marx's theory of the "estranged labour" of capitalism, which "in estranging from man 1) nature, and 2) himself, his own active functions, his life activity estranges the ...*species* from man."[32] However, these established ideas carry new implications in the light of evolutionary psychology and an extra-humanist perspective.

Except for the mostly realistic *Secret Garden*, the books that I examine fall under the definition of speculative fiction. This makes sense, since, as goes without saying, any depiction of humans either living forever or being alive after their death belongs to the realm of speculation rather than of observed reality. Of the two main speculative literary genres, science fiction and fantasy, I have chosen the latter for two reasons. The first is practical: fantasy treatments of human immortality are relatively plentiful in children's literature, while science-fiction treatments of it are almost non-existent—something that may be telling in itself.

The second reason is that the theme of human immortality marks a point of sharp disagreement between the genres of fantasy and science fiction. Author Roger Zelazny explains this disagreement by noting that science fiction "presuppos[es] that people are able to understand and control their lives and surroundings," while fantasy "presupposes human inability to understand or control individual lives or the general environment [and therefore] depicts immortality as an undesirable condition."[33] It is worth noting that this distinction parallels the differing attitudes toward nature that ecocritic Greg Garrard lays out in his categories of "cornucopian" and "Deep Ecology" beliefs, respectively.

While it is true that some fantasy texts portray revenants positively—for example, as angels or spiritual mentors—most of them show a deep revulsion at the idea of human immortality. Carl B. Yoke notes that, despite science-fiction writers' supposed optimism, they as much as anyone tend to depict immortals as "leading boring, meaningless, and lonely existences" and have equal difficulty in envisioning a "life without termination."[34] Intriguingly, Yoke ascribes this to evolutionary biology: "It seems that built into [our] genetic code is a termination point, a program which when complete produces … death. There seems to be a grammar for death built into life itself."[35]

I have identified within my chosen texts four different ways in which treatments of immortality aid the biophilic impulse and serve to "smuggle" nature-oriented values from generation to generation. First, Frances Hodgson Burnett's *The Secret Garden* and Kenneth Grahame's *The Wind in the Willows*—texts that might be categorized with equal validity as pantheistic, animistic, or simply mystical—portray people (or anthropomorphized animals) as being biospiritually integrated within the natural matrix and able to draw on it in mystical ways for their own "eternal life." Secondly, Natalie Babbitt's *The Search for Delicious* and Diana Wynne Jones's *Drowned Ammet* work with animistic belief systems and frameworks of what E. O. Wilson terms "ecological time," in which "individual people and other organisms are no longer distinguishable,"[36] to remind readers that the human species depends for its continued existence on the "immortality" of the earth and its abundance. Thirdly, Ursula K. Le Guin's *The Farthest Shore* and Natalie Babbitt's *Tuck Everlasting* unite characterological and ecological bases in their fierce objection to the modern-day quest for human bodily immortality through life extension technologies. Finally, Le Guin's *The Other Wind* and Philip Pullman's *The Amber Spyglass* present humans' atomistic return to the living universe after their bodily death as a natural and ideal process from which Western religious and scientific values have severed us.

Immortality figures to greater and lesser degrees in these books. It is the central concern of both the Le Guin novels that I analyze, for instance, and more peripheral or intermittent in *The Wind in the Willows*, *The Search for Delicious*, and (arguably) Philip Pullman's His Dark Materials books. I focus my analysis correspondingly more or less narrowly on the theme of immortality in any given book, and frequently pull back to a wider perspective on the context surrounding that theme.

Although, as I mention above, children's literature shows more explicit alarm about pre-death immortalism, and a greater need to push back against it, than the Heavenly afterlife of Christianity, (Judeo-)Christianity also comes in for a rough ride, if more implicitly. This is partly because the authors seem to agree, consciously or not, with Lynn White, Jr., that the Abrahamic belief system, which White calls "the greatest psychic revolution in the history of our culture,"[37] is the main culprit in Western societies' alienation from and exploitation of nature. They also, like White, tend to represent modern Western science as "cast in a matrix of Christian theology."[38]

Therefore, whether overtly or implicitly, my selected texts are fighting the concept of immortality on two fronts, the Christian and the technological, reflecting the extent to which the same arrogation of humans over nature pervades both paradigms. They consistently challenge what White terms Judeo-Christianity's concept of time as "nonrepetitive and linear,"[39] its insistence on "a dualism of man and nature," and its belief that "it is God's will that man exploit nature for his proper ends." Most of them subtly or unsubtly condemn the psychological and socioeconomic extension of this paradigm, the "implicit faith in perpetual progress which was unknown either to Greco-Roman antiquity or the Orient."[40]

At the same time, the novels that I examine in this book, along with (I would argue) most children's books, vehemently reject White's conclusion that "Despite Darwin, we are not, in our hearts, part of the natural process. We are superior to nature, contemptuous of it, willing to use it for our slightest whim."[41] While it may be true (at least for many Americans) that, as Gene Myers remarks, "More than a century after Darwin, we have yet to come to terms with our own animality, including the bonds it gives us, virtually ready made, to other species,"[42] if White's phrase "in our hearts" is meant to indicate something like "in our cellular make-up, in the deepest and most primary impulses of our being," then these authors argue quite the opposite. Most children's literature expresses and nurtures an evolutionarily hardwired understanding that, despite God and Christ and Abraham Darby and Elon Musk, we are, in our hearts, entwined with and fascinated by the natural process, and, to repeat E. O. Wilson's credo, "will find little ultimate meaning apart from [it]."[43]

Biology, ecocriticism, and the humanist echo chamber

In the years immediately after I settled on my topic as that of "ecological perspectives on immortality in children's literature," I referred to my project as being "ecocritical." I soon found it necessary to qualify this designation. As I learned more about ecocriticism and about my own approach, I came to understand that I was working within ecocriticism only in the broadest sense of the term. Specifically, my analysis conforms to its most basic definition, as offered by ecocriticism's founder, Cheryll Glotfelty: it "takes as its subject the interconnections between nature and culture"[44] and studies "the relationship of the human and non-human … entailing critical analysis of the term 'human' itself."[45] Within any narrower bounds than this, though, ecocriticism and I part ways.

My first point of departure from ecocriticism as most academics currently practice it is my dismay at the field's increasing participation in what I think of as the humanist echo chamber. As a student of nature and a longtime aspiring biologist with some formal education and more self-education in the life sciences, I was excited by Glotfelty's claim that "humanities scholars are increasingly making an effort to educate themselves in the sciences and to adopt interdisciplinary approaches."[46] I also hoped that an alliance between the humanities and the more evidence-based fields of the soft and hard sciences would offer an alternative to what I viewed as the obscurantism, unsupported and often unsupportable theories, and illogic that had largely alienated me from the academic humanities during my years at school.

However, these early promises remain largely unfulfilled. The "belligerent attitude to theory"[47] and eschewal of gobbledygook that I found so attractive began a fairly rapid turnaround, one that gained steam after a 2010 Special Forum on Ecocriticism and Theory issue of *Interdisciplinary Studies in Literature and the Environment* (ISLE). In the issue, Serpil Oppermann criticizes ecocritics' "phobia" of theory, arguing that it stems from associating theory with "poststructuralist interrogation of the referents of reality"[48] and saying that to continue to resist incorporating theory as a modus operandi threatened to make "the future of ecocriticism …likely to be one of reactive rollback to dated paradigms."[49] This has opened the door to serious ecocritical consideration of such books such as ecophilosopher Timothy Morton's *Ecology Without Nature*, in which he cites as key influences "Marx and Derrida … Benjamin, Freud, Heidegger, Lacan, Latour, Žižek, and … Hegel," along with judicious applications of Theodor Adorno,[50] and in which he frequently "interrogates the referents of reality."

More importantly, two decades after Glotfelty's claim of interdisciplinarity, Glen Love was still declaring that "serious interdisciplinary work between humanists and the sciences is one of the important tasks that literary ecocritics must take on *in the future*" (emphasis mine).[51] Most of what gets published and discussed has little reference to or participation with the obviously associated fields of biology and evolutionary science, or even the field of ecology from which ecocriticism takes half of its name. This has the effect of perpetuating premises and assumptions that stem from either a Judeo-Christian/Enlightenment paradigm or, no differently, what David Ehrenfeld in *The Arrogance of Humanism* calls "the religion of humanism: a supreme faith in human reason—its ability to confront and solve the many problems that humans face, its ability to rearrange both the world of Nature and the affairs of men and women so that human life will prosper."[52] Meanwhile, few biologists or ecologists have joined in on ecocritical discourse—perhaps because, as Glen Love dryly remarks, "the current diehard constructionism of many literary scholars strikes them as absurd."[53]

In fact, rather than offering rigorous, fact-based, and applied counterbalance to decades of increasingly absurd humanities theorizing, ecocriticism has picked the opposite target, sharing with the humanities "a general coolness, even hostility... toward the sciences in recent decades."[54] Through this attitude, it has indeed suffered a "reactive rollback to dated paradigms." As William Major and Andrew McMurry note, "[E]cocriticism has failed to live up to our expectations, because it begins to look, twenty years or so in, more or less like just another professional discourse among professional discourses."[55]

Leading ecocritic Greg Garrard helps to cement the hollow discursivity of the field by highlighting Glotfelty's statement that "Just as feminist criticism examines language and literature from a gender-conscious perspective, and Marxist criticism brings an awareness of modes of production and economic class to its reading of texts, ecocriticism takes an earth-centered approach to literary studies."[56] Garrard concludes that "[e]cocriticism is, then, an avowedly political mode of analysis, as the comparison with feminism and Marxism suggests," and further notes that "Ecocritics generally tie their cultural analysis explicitly to a 'green' moral and political agenda."[57]

Although this reading of Glotfelty's meaning is questionable and the word "avowedly" an overstatement (or so we can hope), Garrard is undoubtedly right in implying that the field's basic identity is at this point as another stock member of the humanities rather than one conducive to developing an interdisciplinary relationship with the life sciences. Ecocriticism extends in America from the environmentalist awareness and activism of the 1970s and '80s, while in European ecocriticism we see a strong enough left-wing presence, and a sufficient centrality of the Marxist Raymond Williams' 1973 *The Country and the City*, to support Jonathan Bate's assertion that "predominantly in Germany but increasingly in Britain" the political Left has made "the move from red to green."[58] To quote Roman Bartosch and Greg Garrard (the latter in blunter mode), "Ecocriticism developed because ethically-minded or guilt-ridden scholars felt the need to address environmental degradation, species extinction and all the other aspects of our current environmental crisis."[59]

This set of motivations excludes me for a number of reasons. For one, I am neither ethically minded nor guilt-ridden. For two, I find the idea of inserting a "morality" and imposing an "agenda" offensive to the notion of scholarship: "should"-based thinking is inevitably an enemy of reality- and evidence-based thinking. I object to the idea of agenda even more because, as Bartosch and Garrard rightly point out, "it will be the prismatic unpredictability of the classroom that will 'refract' our work [as ecocritics] most beautifully."[60] This is another way of saying that our attitudes toward scholarship will devolve upon our students. That possibility finds expression in various and alarming ways. There is the missionary stance of David Mazel, who demands proof that "students who read and write about green texts turn into more thoughtful and effective environmentalists than they might have been otherwise,"[61] as well as Garrard's call for a long-term study that will show this correspondence (or absence thereof), and the triumphant pronouncements of some Green Studies professors that every semester they "convert" a number of their students to vegetarianism. As I have argued elsewhere, I strongly believe that studying *or* teaching from agenda is a bankrupt concept, one that "traduces and terminally vitiates" the purposes of scholarship and pedagogy.[62]

Still, my personal opinions about the persistence of humanist and Judeo-Christian assumptions within ecocriticism are obviously secondary. It is more to the point that for scholars to remain ignorant of and/or resistant to basic biological, ecological, and evolutionary premises undermines the positivist and interdisciplinary promise of ecocriticism. It also produces some dubious work. Philosopher Kate Soper's much-cited book *What Is Nature?* (1995) demonstrates some of the problems of theorizing about humanity in relationship to nature through a typical humanist/humanities lens, devoid of any grounding in biological or other scientific education. For example, concerning sexuality, Soper writes that "Animals, notably other primates, do indeed manifest sexual hierarchies, and appear to observe certain rules of intercourse, but it is only human beings, who, in virtue of language and conceptualization, can be said to experience themselves as sexual beings, with all the sources of pleasure and pain which that entails."[63]

This is a pronouncement about the awareness and existential experiences of other living beings that neither Soper nor anybody else can possibly support, but it is important to know that the observations and data sets of many zoologists and naturalists would reject it.[64] Few professional observers of animals would subscribe to Soper's position about animal sexuality, language capacity, and possession of "consciousness"—and nor does Soper have the credentials, insofar as anybody does, to make such assertions. Yet she, like many in the culture of the academic humanities, not only makes such assertions freely and frequently, apparently without feeling the need to support them in accordance with academic principles and scholarly standards, but bases a good deal of her argument upon them.

Soper's firmly unscientific work, and the ecocritical community's eagerness to latch onto it, sadly seems to herald ecocritics' increasing absorption into the conventional academic humanities, many of whose members, as Frederick Crews points out, "subscribe to a two-tiered conception of truth [, making] a token bow to empirically grounded knowledge, but…instead of merely avowing that the subjective realm is elusive… advanc[ing] their own preferred theory, which is typically sweeping, absolute, and bristling with partisanship."[65]

Glen Love charitably speculates that "much of this hostility [toward science] is an anachronistic holdover from the wholly justified reactions to the social Darwinist distortions [of sociobiology] a century ago."[66] I agree that the long memory of academe may account for some part of the rejection of the basic precepts of biology, sociobiology and associated evolutionary theories. It seems more likely, though, that (as Love also acknowledges and explores at length) much of the hostility stems simply from a deep-rooted resistance to the idea that human beings are animals and that evolutionary concepts apply to us.

The other major juncture at which I part ways with mainstream ecocriticism is in my focus on what human alienation from nature has done to humans, rather than what it has done to nature. It is fair to say that most current and past ecocriticism puts its focus the other way around. Cheryll Glotfelty, for instance, centers her introduction on "efforts to promulgate environmentally enlightened works"[67] and says that "most ecocritical work shares a common motivation: the troubling awareness that we have reached the age of environmental limits, a time when the consequences of human actions are damaging the planet's basic life support systems."[68] Greg Garrard's lucid primer, *Ecocriticism*, moves from the headings of pollution and "positions" to pastoral and wilderness, but none of this touches on the damage done to *humans* by the past 150 years of industrialism, development, urbanization, and life within "the most anthropocentric religion the world has seen."

Garrard has in fact expressed to me in correspondence and conversation a deep skepticism that there is any particularly detrimental effect on Western humans from our current way of living and thinking. He has cited the historically unprecedented quality of life, life expectancies, and happiness indices within modern-day Western and Westernized countries. Meanwhile, influential American ecocritic Lawrence Buell reaffirms his orientation toward the damage we are doing to nature, rather than what a lack of it is doing to us, with the subtitle of his 2005 book, *The Future of Environmental Criticism: Environmental Crisis and Literary Imagination*.

This suggestion that the Abrahamic, post-Industrialist dispensation has had a sharply asymmetrical impact on nature and humans bespeaks most ecocritics' apparent belief that we humans are, after all, independent of nature, fundamentally unlike other animals, and largely unharmed by our removal from the natural environments within which we evolved over a quarter-million years. Perhaps it is true that our nearly unique level of sociability offsets to a degree the deprivations of living in minimal contact with nature and in conditions that bear little resemblance to those in which we evolved. If so, I do not believe that it is very great a degree. As shown by evidence gathered in current "nature-deficit disorder" studies, it is not a large enough degree to offset the social and psychological effects of being torn between our deep social impulses and our deep biophilic ones.

Here we come back around to my chosen lens of E. O. Wilson's biophilia hypothesis, and hence to sociobiology. The latter is a freighted term, still strongly associated (as Glen Love points out) with the social Darwinism of Herbert Spencer, Arthur Jensen, and, falling chronologically between them, Adolf Hitler. (Since it was Spencer, not Darwin, who coined the term and developed the idea of "the survival of the fittest," the application of this concept to Darwinism is a distortion in the first instance.)

Wilson, in a 1978 defense of attacks against his 1975 book *Sociobiology: The New Synthesis*, explains the term's origin in John P. Scott's suggestion for a more formal way of naming "the interdisciplinary science which lies between the fields of biology (particularly ecology and physiology) and psychology and sociology."[69] He defines the discipline as "the systematic study of the biological basis of all forms of social behavior, including sexual and parenting behavior, in all kinds of organisms including humans."[70]

Wilson notes that his theories met with "stiff resistance."[71] This understates the case to a fabulous degree. In reality, a raft of scientists, philosophers, and graduate students, among them Wilson's illustrious Harvard colleagues Stephen J. Gould and Richard Lewontin, ran a blistering and distortive letter rejecting sociobiology in the November 13, 1975, *New York Review*, and this set the tone of anti-sociobiology arguments thereafter. The group accuses Wilson's book of being "[t]he latest attempt to reinvigorate ... tired theories" of genetic determinism that "provided an important basis for the enactment of sterilization laws and restrictive immigration laws by the United States between 1910 and 1930 and also for the eugenics policies which led to the establishment of gas chambers in Nazi Germany."[72] It is an accusation that Wilson, in a countering letter to the same paper, calls "ugly, irresponsible, and totally false."[73]

The signatories' main contention was that Wilson was exercising expediency in the scope and flexibility of genetic determinism, allowing him to "confirm selectively certain contemporary behavior as adaptive and 'natural' and thereby justify the present social order."[74] Significantly, they also state, "While evolutionary analysis provides a model for interpreting animal behavior, it does not establish any logical connection between behavior patterns in animal and human societies,"[75] phrasing that in my view betrays this group's resistance to the idea—in some quarters considered a fact—that humans are animals.

It should be noted here that Wilson's letter pointed out misquotations and misrepresentations in Gould et al.'s original letter of such an egregious nature that they could hardly have been accidental. For instance, Wilson writes, "The letter signers... allege that I make institutions such as slavery 'seem natural in human societies because of their "universal" existence in the biological kingdom.' I have done no such thing. In the book slavery is stated to occur only in ants and men, and the many distinctions between its practice in the two groups are made clear."[76]

The vitriolic infighting within the scientific community has continued over the past five decades, with British ethologist Richard Dawkins taking over the lead of the anti-Wilson contingent when Stephen Jay Gould died. In this period of time, several factors have led more scientists to at least a nominal acceptance that humans are animals and that to some degree their social and individual behaviors and perceptions derive from evolution. These factors include new data in animal studies, a growth of disciplines dedicated to the study of humans as animals (human behavioral ecology, evolutionary psychology, evolutionary ethics), and, perhaps most significantly, an increase in the use of magnetic-resonance imaging (MRI) techniques to study neurological responses in both humans and other animals.

However, to what degree, at what level of determination, and by what processes humans are an animal species are thorny issues. They have led to a "calving-off" of theories within the wider sociobiological set and, in the words of Louise Barrett, Robin Dunbar and John Lycett, "spawned a large amount of rather acrimonious debates"[77]—particularly between the human behavioral ecology and evolutionary psychology camps—that are nearly as savage as the initial ones about genetic, evolutionary, and biological determinism.

In fact, the original and current debates are in many respects the same one. For instance, proponents of so-called "massive modularity" of the brain argue for an almost unlimited human capacity to adapt to new conditions such as urbanization. By contrast, the believers in general-domain functioning argue that our brains are more fixed. Likewise, the proponents of "environment of evolutionary adaptedness" argue vigorously over whether most of the relevant hominid responses arose in the Pleistocene or if a significant number of evolutionary changes were still occurring in the early Holocene. Meanwhile, many sociobiologists moot many of their findings by employing in their theories the "phenotypic gambit," in which, to use Barrett, Dunbar, and Lycett's words, they "ignore the details of inheritance (genetic or cultural), cognitive mechanisms, and phylogenetic history that may pertain to a given decision role and behavioral domain in the hope that they don't matter in the end result."[78]

To the extent that I feel qualified to comment on these arguments, which is not very far, I perceive in them a number of seemingly false binaries. This leads me to suspect that many of these scientists are operating out of the "dualistic dynamic" that Val Plumwood sees as permeating the Western world view—a dynamic in which a "master class" defines, oppresses, exploits, and denies an underclass, whether that underclass is women or nature. Plumwood believes that overcoming this false dualism involves "recognition of both continuity and difference[:]...acknowledging the other as neither alien to and discontinuous from self nor assimilated to or an extension of self."[79]

Personally, I subscribe on the whole to the "third way" of Robert Boyd and Peter J. Richerson, who propose a "dual inheritance" of the genetic and the cultural and thereby reject the separation of culture and biology "with only simple biological 'constraints' on cultural evolution."[80] The authors point out, "If culture itself has the attributes of an inheritance system, then it makes sense to apply Darwinian analytical methods to [it] as well as to the genetic."[81]

Complications and infighting aside, in this project I use the term "sociobiology" more generally to represent E. O. Wilson's proposition that "[b]eing human simply means being one variety of animal on planet Earth" and that "Our similarities with other species, with whom we share a great deal of our genetic inheritance, turn out to be more remarkable than our differences."[82] I welcome this seemingly self-evident precept both as a platform for viewing child readers as part of the animal kingdom and for the space it provides, however crowded with warring factions, as, to use E. O. Wilson's definition, "the interdisciplinary science which lies between the fields of biology (particularly ecology and physiology) and psychology and sociology."[83]

Social scientist Alexander Rosenberg rightly remarks that Wilson's definition of sociobiology as studying "the biological basis of all forms of social behavior, in all kinds of organisms, including man"[84] does nothing to "distinguish [the] discipline from that of biology *tout court*." He argues that if Wilson merely means that "there is a causal connection between genetic inheritance and each and every event involving the anatomical, physiological, behavioral machinery of the organism," then "[the] claim suffers, not from falsity, but from obviousness."[85] It would be a fine thing indeed if everyone thought that such a connection were too obvious to need stating or defending. But, to return to Val Plumwood's remarks, "in the context of the modern [Western] world," the view of humans as animals and as guided by evolutionary impulses runs so counter to the "dominant strand of thought" as to be not only non-obvious but "revolutionary."

The revolutionary nature of including humans within the scope of Darwinist evolution is perhaps most readily apparent in the United States, where, in a 2010 international assessment of the acceptance of Darwinism, 14 percent of Americans said they believed the theory of evolution was "definitely true"[86]—fewer than the highest national group (the Dutch) who believed it to be "absolutely false." One-third of Americans rejected the theory of evolution out of hand. In a 2012 national poll, 46 percent of Americans believed that God created humans in their current form, and another 32 percent believed that God had guided evolutionary processes to lead to humans in their current form; this left only 15 percent who believed in evolution as a scientific process applied to the human species as well as all other species.

However, as easy as it is to point fingers at the American insistence on creationism, E. O. Wilson notes that the "stiff resistance," not to say virulent backlash, to the basic precepts of sociobiology came chiefly from two extra-national groups. The first was that of social scientists in general, who, in Wilson's words, adhered to "the Durkheim-Boas tradition of the autonomy of the social sciences." The second was that of Marxists, who "cling to a vision of human nature as a relatively unstructured phenomenon swept along by economic forces extraneous to human biology" and whose "secular ideologies, [which] rested secure as unchallenged satrapies of scientific materialism," now "were in danger of being displaced by other, less manageable biological explanations."[87] While this may be an unfair generalization in light of Karl Marx's serious examination of the "species alienation" created by industrial capitalism's mass urbanization, it does go to my earlier point that many groups who ostensibly accept Darwinism are reluctant to step outside the "humanist echo chamber" and perform analyses that genuinely reposition humans within the context of evolutionary biology.

A new way of reading (or: hermeneutics schmermeneutics)

I analyze the texts in my book from the premise that much children's literature simultaneously taps into and fosters in child readers a biological, pre-/non-Christian, non-humanist sense of themselves as members of an animal species interactive and interdependent with their environmental surroundings. By rights, this perspective should be considerably less "alternative" than it is within the 40-year history of academic children's literature criticism. It is true that my approach differs in the ways I mention above from the more "humanist" and political modes of ecocriticism; I picture myself to be on the science-facing edge of the ecocritical sphere, trying to open a section of literary criticism's boundary to the interdisciplinary space between life sciences and social sciences that Wilson originally conceived for sociobiology.

But since, as Dobrin and Kidd note in *Wild Things*, and as is fairly obvious to even the most casual observer, "[C]lassic children's literature has long been preoccupied with natural history, ecology and human-animal interaction,"[88] it came as a surprise to me that the gaps between my position and the substance of children's literature criticism were so wide and so barren. I had expected to be working from a decent fund, if not a wealth, of scholarly and critical work on the nature-based content of children's literature. What I found instead was very little criticism that focused on this aspect, or acknowledged it—or even, in terms of theoretical outlook, allowed it, as I mention at the top. The bulk of existing literature is made up of special "green" issues of the *Children's Literature Association Quarterly* in 1994 and *The Lion and the Unicorn* in 1995; a chapter in Julia Mickenberg's 2005 *Learning from the Left* on the Soviet and American "nature study" movements in children's literature in the 1950s; a somewhat animalistic depiction of the child, derived from Jean-Francois Lyotard's concept of the "monster child," in Kimberley Reynolds' 2007 *Radical Children's Literature*; and Sidney I. Dobrin and Kenneth Kidd's invaluable 2004 compilation *Wild Things: Children's Culture and Ecocriticism*, in which Kidd observes that children's literature criticism and ecocriticism had to that point been "largely separate undertakings."[89]

The short explanation for this disconnection is that, as I have intimated, the two fields consolidated their legitimacy as academic disciplines with very different relationships to the theoretical trends of the day. Both emerged in the academy in the mid- to late 1980s. Modern-day children's literature criticism began, arguably, with the 1984 publication of Jacqueline Rose's *The Case of Peter Pan, or, the Impossibility of Children's Fiction*. Ecocriticism was launched with the adoption at a 1989 Western Literature Association (WLA) conference of the word "ecocriticism" as an umbrella term for disparate nature-oriented analyses, followed by WLA member Cheryll Glotfelty's 1990 appointment as the first American professor of Literature and the Environment. Ecocriticism, seeking status as a meta-outlook that "claims as its hermeneutic horizon nothing short of the literal horizon itself" and mounts a "challenge to most postmodern discourse as well as to the critical systems of the past",[90] opened up "a theoretical space…[that was] largely antipathetical to theory in its then prevalent modality."[91]

Ecocritical scholar Laurence Coupe notes that "[i]n various schools—formalist, psychoanalytic, new historicist, deconstructionist, even Marxist—the common assumption has been that what we call 'nature' exists primarily as a term within a cultural discourse, apart from which it has no being or meaning." He characterizes the then-new discipline as an effort to reclaim an objective reality from the constructionists, to assume that "something is already there, asking to be actualised or understood," in contradistinction to standard critical theory that "seems to have been dedicated to repudiating any such 'realisation'."[92] In the early focus on "how modern conventions of reading block out the environmental dimension of literary texts,"[93] there was, in Rigby and Goodbody's words, a "widespread rejection of ...critical or cultural theory by most first-wave ecocritics."[94] So, at the start, the field developed as a hallmark the "belligerent attitude towards critical theory" that John [95]Parham would decry by 2008.

Children's literature criticism, meanwhile, was knee-deep in suspicion, and wading in farther with every step. In *The Case of Peter Pan*, Jacqueline Rose contends that children's fiction constructs the child within the book as innocent, fixed and "knowable in a direct and unmediated way,"[96] as part of an "often perverse and mostly dishonest" desire on the part of adults to "hold off a panic"[97] about their own sexuality and the unknowable nature of the world. Peter Hunt, writing in 1994, takes a similarly jaundiced view: "the writers and manipulators of children's books are adults; books are makers of meaning for their readers, and their readers are children."[98] Hunt concludes, "[W]hat needs to be emphasized is that children's literature is a powerful literature, and that such power cannot be neutral or innocent, or trivial... . Equally obviously, the primary audience is children, who are less experienced and less educated into their culture than adults. [This] means that [children's books] are part of a complex power-relationship."[99]

Likewise, Canadian scholar Perry Nodelman, who views writers as "workers in a field of economic production"[100] and consistently refers to children's books as "productions," paints the children's novelist as a sinister being, one who by means of double focalization and other techniques inserts a "hidden adult" behind the texts to exploit the power differential and didactically manipulate the child reader. He is convinced, he writes, that "adults have imposed their own theoretical assumptions about children on children—constructed them as the limited creatures the adults have imagined them to be."[101] Karen Coats, working within her psychoanalytic approach with some number of laity-friendly (and even ecocritic-friendly) ideas—for instance, that the "subject" is "split, a construction of both natural and cultural influences, of conscious and unconscious processes"[102]—goes beyond suspicion to a fully paranoiac way of reading when she accepts and applies such Lacanian precepts as that "the child must come under the 'Law of the Father' in order to take up a position in the Symbolic; he must, paradoxically, become his own cause."[103] Perhaps more significantly, she believes that "[s]ince our encounters with difference are always linked to the grief of separation from the mother, our imaginary constructions of race and gender are fraught with sadness and fear."[104]

Of course, classical psychoanalysis, as a field that studies humans *qua* humans, is by definition one of the most highly anthropocentric lenses through which a subject can be viewed. While ecocriticism (at least theoretically) admits the idea of human as animal and claims an interest in expressions of, and possibilities for, our existential reintegration with the non-human world, most lines of psychoanalysis and its related fields focus on the child's purported process of learning that he or she is separate from and "above" the merely biological and animalistic. This has created in children's literature criticism a historically biophobic culture. It not only puts a disproportionate focus on psychoanalysis-amenable texts that are at least as anomalous as they are canonical—most signally the acid-trippy *Peter Pan* and *Alice in Wonderland*—but a tendency to "problematize" the idea of child-nature affinity from a position of dismissal, if not contempt.

For instance, Bettelheim's *The Uses of Enchantment*, waning gradually in influence but still cited by hundreds of scholars and widely used in university courses, reads child-animal bonding in fairy tales and fantasies not as archetypal expressions of humans' inherent connection with other members of the living world, but as an indication of the "pre-rational" state of the child, who is still (Bettelheim notes pityingly) ignorant enough to believe that "whatever has life has life very much like our own."[105] It has also led to tendentious-seeming and perhaps even willfully anti-ecological readings of key texts in children's literature criticism, such as Karen Coats' reading of *Charlotte's Web* that I cite at the top.

Where children's literature scholars followed the more general critical trends, they ran equally afoul of an ecocritical mandate to have natural elements in texts read "as a …literal reference or as an object of retrieval or contemplation for its own sake" rather than for "its formal or symbolic or ideological properties."[106] Theoretical frameworks such as deconstructionism and post-structuralism (see, for example, Karen Lesnik-Oberstein's 1994 *Children's Literature: Criticism and the Fictional Child*), with their "radical textualism,"[107] propound, in Kate Soper's words, an "irreducibly cultural and symbolic order of human being," while they also "consistently criticized naturalist explanations of the being of humanity," and looked on the idea of inherent value in nature "with suspicion…as an attempt to 'eternize' what in reality is merely conventional."[108]

Many aspects of psychoanalytic literary theory and other modes of "suspicious" critique keep them separate from ecocritical (or at least early ecocritical) possibilities. Firstly, they discourage critics from considering in any kind of face-value way the causes, functions, and effects of the strong presence of nature in children's literature. Secondly, they forestall the recognition that, as Gene Myers puts it, "as a child's social abilities develop, he or she perceives an animal as another being with subjective experience of its own."[109] They also agitate against the commonsensical perception that a strong identification with nature results not from a person's life-stage, but from "the innately emotional affiliation of human beings to other living organisms," as Wilson puts it in his biophilia hypothesis.[110]

Instead, Myers suggests, those engaged in traditional theories of child development and the literature that forms part of it reject the idea that animals are crucial to human life, and instead "have dispersed them to the official domains of child psychology: here in conceptual development, a bat that is not a bird; over there in psychoanalysis, the horse that is the father. Encounter is mediated, indirect, because some more important human feature intervenes."[111] Myers goes on to observe that historically characterizing interactions between children and animals (and, by extension, other aspects of the biotic world) from within this set of traditional humanist perspectives practically guarantees that the nature-based content of children's texts will be read for its psychoanalytical possibilities.

> Some, influenced by psychoanalytic traditions, will argue that the child's interactions with the animal are actually projective episodes, in which some intrapsychic conflict is acted out on the animal, with indifference to its distinct being. Others (though this position is weakening) are likely to object that the child's cognitive apparatus is too egocentric to differentiate the animal as a nonanthropomorphic other; instead, the child assimilates the animal to the human pattern. Still others...hold that language and culture are determinative over all else and that the meanings of the animal to the child are explained by such human-generated sources.[112]

Despite the foregoing, it would seem on paper that no number of theoretical incompatibilities could keep ecocriticism and children's literature criticism wholly apart, for their fields of analysis share a common progenitor: the Romantics. Humphrey Carpenter proposes, and most critics agree, that the modern genre of children's literature—that is, the post-tractarian trend that began in the so-called "Golden Age" of Victorian and Edwardian children's books[113]—emerged from "a change... in the attitude of adults to children, a change closely bound up with the Romantic movement."[114] Carpenter views Wordsworth's "Ode" and William Blake's Songs of Innocence as "a call to revolution against the view of childhood [in which]...a child was simply a miniature adult, a chrysalis from which a fully rational and moral being would duly emerge,"[115] and toward the idea that "children are in a higher state of spiritual perception than adults, because of their nearness to their birth."[116] But, Carpenter argues, although imaginative writing for children was viable from around 1830 on, books such as *Alice in Wonderland*, *The Secret Garden*, and *The Wind in the Willows* didn't emerge until the later 1800s and early 1900s, because it was then that "[t]he mid-Victorian belief in progress…was starting to be shaken"[117] and the squalor and miseries of Industrialism began to come clear.[118]

Meanwhile, what Jonathan Bate calls the "received wisdom"[119] and commonsense view of the Romantics themselves is that they were nature writers. "[I]f the French Revolution was one great root of Romanticism," he writes, "'the return to nature,' associated above all with the Rousseau of the second *Discourse* and the *Nouvelle Heloise*, was surely the other."[120] In the late 18th and early 19th centuries, the Romantic writers and artists presented "a new narrative [which] came to challenge the earlier story of divine order"[121] and gave rise to a "view of the interconnectedness of all living—and even non-living—things."[122]

All of this would be fine and well, and ecocriticism and children's literature surely prepared for an ardent sibling embrace, were it not for the infiltration of Romantic Studies themselves by that same "hermeneutic of suspicion" that bedeviled children's literature criticism. As Bate outlines, the "Victorian way of reading" the Romantics—that is, chiefly as nature writers, though also recognizing that political events, particularly the French Revolution, played a major role—gave way in the mid-20th century to "idealizing, imagination-privileging critics" such as Geoffrey Hartman and Harold Bloom."[123] These influential figures strove to bring philosophy to the center of Romantic readings, and, following Paul de Man, argued that poets such as Wordsworth placed so much premium on the imaginative faculty as to suggest that we do not need nature: we can create it, in effect, through our mind's eye.

Things got no better for a nature-centered view of the Romantics with the New Historicism of Jerome McGann. Greg Garrard,[124] performing a brief but persuasive rehabilitation of pastoral as a "radical" form of literature, cites historicism as "the critical hegemony that would most object to the [idea of radical pastoral], that is most hostile to ecocriticism, and that nevertheless seems its *sine qua non*."[125] He cites ecocritic Karl Kroeber as observing that McGann imported into Romantic studies a theoretical lens "that extended Cold War psychology into academe and cemented assumptions of the superiority of the critic over the artist."[126] This approach, as Garrard notes, allowed for a vision of the Romantic poets and other ecocentric artists as "collaborationist" and "complicit" with repressing "true history."[127] As Bate sums things up, "Hartman threw out nature to bring us the transcendent imagination; McGann throws out the transcendent imagination to bring us history and society."[128]

The end result is that ecocritics (starting particularly with Bate) had considerable work to do before they could "find English and American romanticism congenial surroundings from which to launch a coherent attack on the critical orthodoxies of today."[129] Children's literature critics would need to perform something of the same rehabilitation to restore the Romantics as the common denominator between the two critical approaches that they would naturally seem to be.

But what about ecocriticism? Why should it be the task of children's literature criticism to accommodate an ecocritical view, and not the task of ecocritics to explore the immense trove of material in the realm of children's literature?

To me, this is the more perplexing cause of the gap between the two fields. Ecocriticism is nothing if not an expansive and accommodating critical view, yet it has featured children's literature even less than children's literature criticism has featured ecocritical views. For instance, *ISLE*, the official journal of the Association for the Study of Literature and the Environment (ASLE), publishes only the occasional article on children's literature, and these treat marginalized and/or obscure texts rather than opening the door to the wealth of contemporary and canonical children's bestsellers. Further, none of the major ecocritical anthologies published to date—including Glotfelty and Fromm's 1996 *Ecocritical Reader*, Laurence Coupe's 2000 *Green Studies Reader*; Armbruster and Wallace's 2001 second-wave compilation *Beyond Nature Writing*; LeMenager, Shrewry, and Hiltner's 2011 *Environmental Criticism for the Twenty-First Century*; Hubert Zapf's 2016 *Handbook of Ecocriticism and Cultural Ecology*; and Aurelie Chone's 2017 *Rethinking Nature*—mentions children's literature at all, much less dedicates essays to it.

Clearly those of us hoping to advance either ecocritical approaches within the field of children's literature or children's literature within the field of ecocriticism face something of a challenge, and one that makes establishing an ecocritical children's canon a more effortful enterprise than one might think. Sidney Dobrin and Kenneth Kidd identify some children's texts as "challenging the consensus understanding of ecocriticism,"[130] but even the most humanist forms of ecocriticism look likely to have a harder time challenging the "consensus understanding" of children's-literature studies. This is doubly true for a biological and evolutionary viewpoint such as the one I take in this book.

There are, however, some hopeful signs. Kidd, Dobrin, Jane Darcy, and Nathalie Op de Beeck have done consistent work in examining children's texts through ecological viewpoints. The MLA sponsored a forum on "Ecocriticism and Young Readers" at its 2012 convention. And, in turn, the 2013 biennial ASLE conference included several children's literature–focused seminars: Companions in Wonder: Children and Adults Exploring Nature Together; Revising Place-Conscious Composition; and Sense of Nature and Ecopedagogical Design in Swedish and German Children's Literature.

As the ASLE conference titles suggest, what points of intersection do currently exist between the two fields of criticism rely on the humanities-based mode of ecocriticism that I describe (and criticize) earlier. It would certainly seem to be the case that so far, formal children's literature criticism has remained almost untouched by extradisciplinary ideas that accord with the genre's nature-oriented content. These include the biophilia hypothesis, which states that humans are "ultimately obedient to biological law";[131] ecopsychology, which "overtly acknowledges and promotes our biological connection with nature as a therapeutic recovery from our misguided detachment from earth";[132] and descriptive evolutionary ethics, a largely empirical and explanatory discipline that examines human and non-human moral behavior vis-à-vis its possible evolutionary survival functions. Time will tell if my own approach has the potential to extend the boundaries of each field to a point of overlap, or if they will remain for all intents and purposes incommensurable.

To read children's texts through a (socio)biological lens and from the premise that many of its authors are writing from a similar outlook has a surprisingly strong effect on the debate over the social and psychological role of children's fiction, a central issue about which children's literature critics have disagreed sharply.

Kimberley Reynolds, revisiting Jacqueline Rose's thesis two decades later, notes that some highly regarded scholars of children's literature, such as Jack Zipes and Robyn McCallum, have joined with Rose in regarding children's fiction as "implicated in acculturation—in transmitting cultural values and 'civilizing' children."[133] Reynolds herself, however, points out that while children's texts are indeed remarkably effective at "filling the minds of generations of young readers with experiences, emotions and the mental furniture and tools necessary for thinking about themselves and the world they inhabit," this affective ability does not necessarily give children's literature "an innately conservative effect":[134] instead, it may make the genre "replete with radical potential."[135]

If we read from the view that much children's literature both depends on and fosters the primal trait of biophilia, these kaleidoscopic critical fragments fall into a coherent pattern. It may be true that many of the genre's texts are "implicated in acculturation." However, if we admit the biophilic premise, then the cultural values in question are often ancient, ineradicable ones that work *against* many modern-day values of the adult Western world—ones that, in Glen Love's words, stem "not just [from] culturally coded determinants but also [from] natural determinants that antedate these, and will outlast them."[136] They are values that perhaps came most explicitly into the modern sensibility with the Romantics but have found expression in many ways, even through the Christian millennia and even in the most repressive times.

In other words, if, as the conspicuous abundance of nature-based texts suggests, one of the chief memes that children's literature transmits and conserves is the human impulse to stay oriented toward and affiliated with other living things and the natural world, then the genre is revolutionary by the very fact of being conservative, and radical in maintaining an otherwise sublimated and disavowed orthodoxy. My book therefore offers an alternative critical perspective that has the potential to unite some of the seemingly conflicting scholarly perceptions of the field.

Children's literature as a Trojan Horse

Lisa Lebduska illuminates a peculiarity of children's literature when she remarks of Dr Seuss's *The Lorax* that extending or applying its ecological message would lead to nothing less than "a reexamination of American lifestyles."[137] In my view, the same could be said of many children's texts. The repercussions would not stop at re-examination, nor at mere lifestyle, nor yet at Americans: they would entail a revolution in the Western world of species self-perception, value systems, priorities and behaviors. Within the eight main texts in this book alone, for instance, reside environmental themes and theses that directly contest many precepts of Christianity and Judaism, strongly criticize industrialism and technological progress, assail equally the "single-visioned" Baconian science of the Enlightenment[138] and the arrogance of humanism, decry the effects of industrial capitalism, present alternatives to Western conceptions of individuality, depict the notion of Manifest Destiny as a social and environmental disaster, exalt ritualized nature-worship, make persuasive cases for the superiority of Greco-Roman and pre-Christian Celtic belief systems, and offer fully realized syncretic spiritual alternatives to the prevailing Western social and religious orthodoxies.

Yet although the guardians of children's literature have been vigilant for "unsuitable" content within certain obvious categories (being chiefly sexuality, death, and religion), they have let these texts, which contain radical and potentially radicalizing ecological/existential messages, slip through the net. Considering that most readers of children's chapter books have not yet reached the age of purchasing independence and many readers of picture books often have not reached the age of reading independence, there is a case to be made that the protectionist phalanx of parents, librarians, teachers, publishers, and religious leaders must have been actively participating in transmitting these subversive texts down through the generations, despite surely knowing what many of them imply.

How do we account for this, as well as for the overwhelmingly nature-based content of Western children's literature in general? We could, of course, ascribe it to simple oversight, but this is to beg the question, since the reason for the oversight is the issue at hand. It is my contention that the publishing history of children's literature since the late 1800s in Britain and the early 1900s in North America reflects and provides a vehicle for our culture's biophilic impulses. Far from being the agent of conformism that many early and late children's literature critics have proposed, the ecocentric core of the generic canon from the Golden Age onward has acted, I believe, as an agent of social subversion, keeping intact pre-Industrial, non-Abrahamic values and visions regarding our surrounding environment. It may even, as I argue in my first chapter, help psychological and cultural expressions of biophilia form a kind of protective layer within the child reader's mind that keeps biophobic beliefs from fully taking hold.

If this premise is granted, it follows that the typical child reader becomes quite a conflicted adult. A key element of my argument is that the "Trojan Horse" function of children's literature—that is, its role in smuggling subversive values past a (deliberately?) blind spot of adult gatekeepers—by the same token fosters a sharp break in the Western reader's value system. Once the reader reaches young adulthood, the nature-oriented perspective, impulses, and allegiances fostered by children's literature are replaced by indoctrination into an entirely opposing orthodoxy. This Western, capitalist, Judeo-Christian adult orthodoxy vigorously promotes a sense of human exemption and exceptionalism, models of lifestyle and ambition that can only be achieved through the exploitation of nature, progressivist ideas that seek to take the human population ever further away from its animal identity, and so on. The former child reader is taught to disdain, exploit, and arrogate themselves above the same natural world that they were originally told they belonged to, and with which they have been encouraged to strongly identify.

I refer to this sharp split as "internal anomie," and I explore it particularly in my chapter on Diana Wynne Jones's *Drowned Ammet* and Natalie Babbitt's *The Search for Delicious*. These books are valuable in having their protagonists cross from the child paradigm/value set to the adult one and back again, exposing the stark and painful conflict between them.

Remarkably few children's books do this. In fact, despite its overarching tendency to foster love and reverence for nature, the genre maintains a conspicuous silence about the problematic relationship between these feelings and Western society's economic, cultural, and spiritual dispensation. Although some canonical children's books, such as *Charlotte's Web* and *King of the Wind*, hint at this conflict by featuring adult characters with a much more utilitarian and domineering view of animals and nature, far more of them simply stay immersed in a childhood world and avoid the forced repudiation of values ahead.

For example, dozens (if not hundreds) of "rockhound" and "young geologist" books published for pre-teens encourage children in a love for banging rocks open with a hammer, but avoid telling them that the most likely occupational outlet for this passion in the wage-earning world will be as a geological engineer for an oil company or strip-mining firm. Legions of both non-fiction and fiction books foster children's fascination with trees, flowers, moss, and other plants without mentioning that their main option for expressing this love in the adult world will be as a monoculture farmer or logger. The animal-lover does not get to the part of the story where she or he can only find work as a lab biologist performing vivisectionist experiments; the children's books portraying the nature-filled idyll of rural living rarely mention that there are few jobs available in these communities and the high-school graduate will almost certainly be obliged to move to the city.

Thus, the "happy endings" ascribed to the genre of children's literature books might be more accurately described as the absence of an ending—the tale of an intense relationship with the rest of the living world that stops just before the divorce. Although my book focuses on the process by which the childhood end of this internal break is fostered, I feel that the implications and extensions are worth considering.

A note on terminology

My diction in this work suffers from a paucity of available terms to signify the natural world. I use terms such as "ecology," "environment," "nature," "the biotic/abiotic matrix," and "the living world" with scandalous interchangeability simply to avoid repeating myself. I recognize that each of these terms represents a separate and sometimes quite different thing, and I only wish I could use the word "creation" to convey what most of the time I really mean. Sadly, I feel that this would run me into some trouble in literal-minded quarters.

Likewise, I occasionally swap the word "ecocentric" for Kate Soper's term "nature-endorsing," now that the former term has apparently fallen out of favor among the cognoscenti. That said, I find "nature-endorsing" inelegant and imprecise, so most of the time I go ahead and use the old-fashioned "ecocentric" or my own clunky term "nature-oriented."

Notes

1 Jonathan Bate, *Romantic Ecology* (Routledge, [1991] 2013), 4.
2 Lawrence Buell, *The Environmental Imagination: Thoreau, Nature Writing, and the Formation of American Culture* (Harvard University Press, 1995), 6–8.
3 Kenneth B. Kidd, *Freud in Oz: At the Intersections of Psychoanalysis and Children's Literature* (University of Minnesota Press, 2011), viii.
4 Kenneth B. Kidd, "A is for Auschwitz: Psychoanalysis, Trauma Theory, and the 'Children's Literature of Atrocity,'" *Children's Literature* 33 (2005): 164.

5 Kidd, *Freud*, viii.
6 Kidd, *Freud*, viii.
7 Bate, *Romantic Ecology*, 6.
8 Diane, Roback, ed., "All-Time Bestselling Children's Books," *Publisher's Weekly* 248, no. 51 (December 17, 2001): n.p.
9 Karen Coats, *Looking Glasses and Neverlands: Lacan, Desire, and Subjectivity in Children's Literature* (University of Iowa Press, 2004), 18.
10 E. B. White, *Letters of E. B. White* (HarperCollins, [1976] 2006), 325.
11 White, *Letters*, 333.
12 White, *Letters*, 543.
13 Sidney I. Dobrin and Kenneth B. Kidd, eds., *Wild Things: Children's Culture and Ecocriticism* (Wayne State University Press, 2004), 233.
14 David J. Buller, *Adapting Minds: Evolutionary Psychology and the Persistent Quest for Human Nature* (MIT Press, 2005), 422.
15 E. O. Wilson, *Biophilia* (Harvard University Press, 1984), 130.
16 Wilson, *Biophilia*, 80–81.
17 Wilson, *Biophilia*, 1.
18 Wilson, *Biophilia*, 85.
19 Wilson, *Biophilia*, 101.
20 Olin Eugene, Jr. (Gene) Myers and Carol D. Saunders, "Animals as Links toward Developing Relationships with the Natural World," in *Children and Nature: Psychological, Sociocultural, and Evolutionary Investigations*, eds. Peter H. Kahn and Stephen R. Kellert (MIT Press, 2002), 154.
21 Val Plumwood, *Feminism and the Mastery of Nature* (Routledge, 1993), 6.
22 Plumwood, *Feminism and Mastery*, 6.
23 Lynn White, Jr., "The Historical Roots of Our Ecologic Crisis," in *The Ecocriticism Reader: Landmarks in Literary Ecology*, eds. Cheryll Glotfelty and Harold Fromm (University of Georgia Press 1996), 9.
24 White, Jr., "Historical Roots," 10.
25 J. Robert Adams, *Prospects for Immortality: A Sensible Search for Life After Death* (Baywood, 2003), 2.
26 Simon Tugwell, *Human Immortality and the Redemption of Death* (Templegate, 1990), 5.
27 John Collier, *Indians of the Americas: The Long Hope* (Mentor Books, [1947] 1960), 7.
28 Kate Soper, *What Is Nature? Culture, Politics, and the Non-Human* (Blackwell, 1995), 8.
29 Kimberly Reynolds, *Radical Children's Literature: Future Vision and Aesthetic Transformations in Juvenile Fiction* (Palgrave Macmillan, 2007), 3.
30 Jean-Jacques Rousseau, *Emile: Or, On Education*. 1762. (Floating Press, 2009), 13.
31 William Wordsworth, "Ode: Intimations of Immortality," in *Major British Poets of the Romantic Period*, ed. William Heath (Macmillan, 1975), 67–69, 257–59.
32 Karl Marx, *Economic and Philosophic Manuscripts of 1844*, 75, accessed May 2, 2013, https://www.marxists.org.
33 Roger Zelazny, "Why Was I Born If It Wasn't Forever?" in *Death and the Serpent: Immortality in Science Fiction and Fantasy*, eds. Carl Yoke and Donald M. Hassler (Greenwood, 1985), 135.
34 Yoke, "Why Was I Born?" 15.
35 Yoke, "Why Was I Born?" 15.
36 Wilson, *Biophilia*, 43.

37 White, Jr., "Historical Roots," 9.
38 White, Jr., "Historical Roots," 11.
39 White, Jr., "Historical Roots," 9.
40 White, Jr., "Historical Roots," 9.
41 White, Jr., "Historical Roots," 12.
42 O. Gene Myers, *The Significance of Children and Animals: Social Development and Our Connection to Other Species* (1997. Rev. ed. Purdue University, 2007), xiv.
43 Wilson, *Biophilia*, 81.
44 Cheryll Glotfelty, "Introduction: Literary Studies in an Age of Environmental Crisis," in *The Ecocriticism Reader: Landmarks in Literary Ecology*, eds. Cheryll Glotfelty and Harold Fromm (University of Georgia Press, 1996), xix.
45 Greg Garrard, *Ecocriticism* (Routledge, 2004), 5.
46 Glotfelty, "Introduction," xxii.
47 John Parham, qtd in Serpil Oppermann, "Ecocriticism's Phobic Relations with Theory," *ISLE* 17, no. 4 (2010), 768.
48 Serpil Oppermann, "Ecocriticism's Phobic Relations with Theory," *ISLE* 17, no. 4 (2010), 768.
49 Oppermann, "Ecocriticism's Phobic Relations," 768.
50 Timothy Morton, *Ecology without Nature: Rethinking Environmental Aesthetics* (Harvard University Press, 2007), 7.
51 Glen A. Love, *Practical Ecocriticism: Literature, Biology, and the Environment* (University of Virginia Press, 2003), 47.
52 David Ehrenfeld, *The Arrogance of Humanism* (Oxford University Press, 1978), 5.
53 Love, *Practical Ecocriticism*, 46.
54 Love, *Practical Ecocriticism*, 6.
55 Greg Garrard, foreword, in *Teaching Environments*, eds. Sieglinde Grimm and Roman Bartosch (Peter Lang, 2014), 231. It should be noted that these two writers, tangling in an extended exchange with Greg Garrard, me, and Roman Bartosch over the purpose of a 2012 symposium at the University of Cologne (and, in a larger sense, over the purpose of ecocriticism in general), speak for me but from the opposition direction: they are lamenting the lack of activism, where I lament its presence.
56 Glotfelty, "Introduction," xviii.
57 Garrard, *Ecocriticism*, 3.
58 Bate, *Romantic Ecology*, 10.
59 Bartosch and Garrard, "The Function of Criticism," 221.
60 Bartosch and Garrard, "The Function of Criticism," 220.
61 David Mazel, "Ecocriticism as Praxis," in *Teaching North American Environmental Literature*, eds. Laird Christensen, Mark C. Long, and Fred Waage (MLA, 2008), 42.
62 Swanigan, "The Case Against Agenda," 237.
63 Soper, *What Is Nature?* 125.
64 Many species demonstrate complex levels of awareness, non-biological patterns of sexuality, ritualized and symbolic behavior, and so on. The notion that only the human species possesses language and can use it symbolically is one that even a moderately educated layperson such as I can refute. The waggle dance of the honeybee uses symbolic language to direct other bees to food sources; observers of wild turkeys have noted at least 30 dialects within their language. This is not even to address the enormous realm of olfactory communication, with its innate ability to convey the subtlest emotional, psychologi-

cal, and physical information, to which most mammals other than humans are privy.
65 Frederick Crews, *Follies of the Wise: Dissenting Views* (Shoemaker Hoard, 2006), 9.
66 Love, *Practical Ecocriticism*, 6.
67 Glotfelty, "Introduction," xxiii.
68 Glotfelty, "Introduction," xx.
69 E. O. Wilson, "What Is Sociobiology?" *Society* 15, no. 6 (1978): 10.
70 Wilson, "What Is Sociobiology?" 10.
71 Wilson, "What Is Sociobiology?" 10.
72 J. Beckwith et al., "Against 'Sociobiology,'" *New York Review of Books*, November 13, 1975, 43.
73 E. O. Wilson, "For Sociobiology," *New York Review of Books*, December 11, 1975, n.p, http://www.nybooks.com/articles/archives/1975/nov/13/against-sociobiology.
74 Beckwith et al., "Against 'Sociobiology'," 43.
75 Beckwith et al., "Against 'Sociobiology'," 44.
76 Wilson, "For Sociobiology," n.p.
77 Louise Barrett, Robin Dunbar, and John Lycett, *Human Evolutionary Psychology* (Oxford University Press, 2002), 10.
78 Barrett, Dunbar, and Lycett, *Evolutionary Psychology*, 10.
79 Plumwood, *Feminism and Mastery*, 6.
80 Peter J. Richerson and Robert Boyd, "Culture is Part of Human Biology: Why the Superorganic Concept Serves the Human Sciences Badly," Draft 3 (7.17.01) in *Science Studies: Probing the Dynamics of Scientific Knowledge*, eds. Sabine Maasen and Matthias Winterhager (Transcript Verlag, 2001), 2.
81 Richerson and Boyd, "Culture is Part of Human Biology," 4.
82 John Gribbin and Mary Gribbin, *The One Per Cent Advantage: The Sociobiology of Being Human* (Blackwell, 1988), 1.
83 Wilson, "What Is Sociobiology?" 10.
84 Wilson, "What Is Sociobiology?" 10.
85 Alexander Rosenberg, *Sociobiology and the Preemption of Social Science* (Johns Hopkins University Press, 1980), 155.
86 *Science*, Oct 2010.
87 Wilson, "What Is Sociobiology?", 10.
88 Dobrin and Kidd, *Wild Things*, 4.
89 Dobrin and Kidd, *Wild Things*, 3.
90 Love, *Practical Ecocriticism*, 1.
91 Kate Rigby and Axel Goodbody, "Introduction," in *Ecocritical Theory: New European Approaches*, ed. Axel Goodbody (University of Virginia Press, 2011), 2.
92 Laurence Coupe, "Introduction," in *The Green Studies Reader: From Romanticism to Ecocriticism*, Laurence Coupe (Routledge, 2000), 2.
93 Buell, *Environmental Imagination*, 14.
94 Rigby and Goodbody, "Introduction," 1.
95 John Parham, "The Poverty of Ecocritical Theory: E.P. Thompson and the British Perspective," *New Formations: A Journal of Culture, Theory and Politics* 64 (2008): 25.
96 Jacqueline Rose, *The Case of Peter Pan, or, The Impossibility of Children's Fiction* (Macmillan, 1984), 9.
97 Rose, *Peter Pan*, 10.
98 Peter Hunt, *An Introduction to Children's Literature* (Oxford University Press, 1994), 2.
99 Hunt, *An Introduction*, 3–4.

100 Perry Nodelman, *The Hidden Adult: Defining Children's Literature* (Johns Hopkins University Press, 2008), 4.
101 Nodelman, *Hidden Adult*, 85.
102 Coats, *Looking Glasses*, 6.
103 Coats, *Looking Glasses*, 7.
104 Coats, *Looking Glasses*, 8.
105 Bruno Bettelheim, *The Uses of Enchantment: The Meaning and Importance of Fairy Tales* (Knopf, 1991), 46.
106 Buell, *Environmental Imagination*, 85.
107 Garrard, "Radical Pastoral?" 183.
108 Soper, *What Is Nature?*, 6.
109 Myers, "Animals as Links," 154.
110 Stephen R. Kellert and E. O. Wilson, eds., *The Biophilia Hypothesis* (Island Press, 1995), 31.
111 Myers, *Children and Animals*, xiv.
112 Myers, *Children and Animals*, 8.
113 Humphrey Carpenter, *Secret Gardens: A Study of the Golden Age of Children's Literature* (Faber, 1985), 2.
114 Carpenter, *Secret Gardens*, 14.
115 Carpenter, *Secret Gardens*, 16.
116 Carpenter, *Secret Gardens*, ix.
117 Carpenter, *Secret Gardens*, 21.
118 Where Carpenter specifies a turn from Lockean to Rousseauian beliefs about childhood and education, Jacqueline Rose dates modern children's literature from "a time when the concept of childhood was dominated by the philosophical writings of Locke and Rousseau." (8) We can perhaps tell, however, from the common yoking of two thinkers who in their attitudes toward nature sat at opposing poles—Locke an empiricist and one of the chief agents of Enlightenment thinking; Rousseau not only a leading figure in early Romanticism but a proto-deep ecologist and evolutionary ethicist—the degree to which ecological viewpoints have been neglected.
119 Bate, *Romantic Ecology*, 7.
120 Bate, *Romantic Ecology*, 8.
121 Ashton Nicols, *Beyond Romantic Ecocriticism: Toward Urbanatural Roosting* (Palgrave Macmillan, 2011), xvi.
122 Nicols, *Beyond Romantic Ecocriticism*, xvi.
123 Bate, *Romantic Ecology*, 6.
124 Just to note: Jonathan Bate was Greg Garrard's Ph.D. supervisor.
125 Garrard, "Radical Pastoral?" 182.
126 Garrard, "Radical," 182–83.
127 Garrard, "Radical," 183.
128 Bate, *Romantic Ecology*, 8.
129 Garrard, "Radical," 182.
130 Dobrin and Kidd, "Introduction: Into the Wild," *Wild Things*, 12.
131 Kellert and Wilson, *Biophilia Hypothesis*, 80–81.
132 Bob Henderson et al., "Pedagogical Response to *The Lorax*," in *Wild Things: Children's Culture and Ecocriticism*, eds. Sidney I. Dobrin and Kenneth Kidd, 137. Wayne State University Press, 2004.
133 Reynolds, *Radical Children's Literature*, 2.
134 Reynolds, *Radical Children's Literature*, 3.
135 Reynolds, *Radical Children's Literature*, 2.
136 Love, *Practical Ecocriticism*, 1.

137 Lisa Lebduska, "Rethinking Human Need: Seuss's *The Lorax*," *Children's Literature Association Quarterly* 19 (1994–1995): 170–76. Qtd in Bob Henderson et al., "Pedagogical Response to *The Lorax,*" in *Wild Things: Children's Culture and Ecocriticism*, eds. Sidney I. Dobrin and Kenneth Kidd, 142. Wayne State University Press, 2004.
138 Roszak, "Against Single Vision," 21–22.

Bibliography

Adams, J. Robert. *Prospects for Immortality: A Sensible Search for Life After Death*. Death, Value and Meaning series. Edited by John D. Morgan. Baywood, 2003.
Barrett, Louise, Robin Dunbar, and John Lycett. *Human Evolutionary Psychology*. Oxford University Press, 2002.
Bartosch, Roman, and Greg Garrard. "The Function of Criticism. A Response to William Major and Andrew McMurry's Editorial." In *Teaching Environments: Ecocritical Encounters*, edited by Roman Bartosch and Sieglinde Grimm. Frankfurt: Peter Lang, 2014.
Bate, Jonathan. *Romantic Ecology: Wordsworth and the Environmental Tradition*. 1991. Routledge, 2013.
Beckwith, J., et al al. "Against Sociobiology." *New York Review of Books*, 22 (November 13, 1975): 43–44.
Bettelheim, Bruno. *The Uses of Enchantment: The Meaning and Importance of Fairy Tales*. Knopf, 1991.
Buell, Lawrence. *The Environmental Imagination: Thoreau, Nature Writing, and the Formation of American Culture*. Belknap/Harvard University Press, 1995.
Buell, Lawrence. *The Future of Environmental Criticism*. Blackwell, 2005.
Buller, David J. *Adapting Minds: Evolutionary Psychology and the Persistent Quest for Human Nature*. MIT Press, 2005.
Carpenter, Humphrey. *Secret Gardens: A Study of the Golden Age of Children's Literature*. Faber, 1985.
Coats, Karen. *Looking Glasses and Neverlands: Lacan, Desire, and Subjectivity in Children's Literature*. University of Iowa Press, 2004.
Collier, John. *Indians of the Americas: The Long Hope*. Mentor Books, 1947.
Coupe, Laurence, ed. *The Green Studies Reader: From Romanticism to Ecocriticism*. Routledge, 2000.
Crews, Frederick. *Follies of the Wise: Dissenting Views*. Shoemaker Hoard, 2006.
Dobrin, Sidney I., and Kenneth Kidd, eds. *Wild Things: Children's Culture and Ecocriticism*. Wayne State University Press, 2004.
Ehrenfeld, David. *The Arrogance of Humanism*. Oxford University Press, 1978.
Garrard, Greg. *Ecocriticism*. Routledge, 2004.
Garrard, Greg. "Radical Pastoral?" In *The Green Studies Reader: From Romanticism to Ecocriticism*, edited by Laurence Coupe. Routledge, 2000.
Glotfelty, Cheryll. "Introduction: Literary Studies in an Age of Environmental Crisis." In *The Ecocriticism Reader: Landmarks in Literary Ecology*, edited by Cheryll Glotfelty and Harold Fromm. University of Georgia Press, 1996.
Glotfelty, Cheryll, and Harold Fromm, eds. *The Ecocriticism Reader: Landmarks in Literary Ecology*. University of Georgia Press, 1996.
Gould, Stephen J., et al. "The Politics of Sociobiology." In *The New York Review of Books*, May 31, 1979. http://www.nybooks.com/articles/archives/1979/may/31/the-politics-of-sociobiology/

Gribbin, John, and Mary Gribbin. *The One Per Cent Advantage: The Sociobiology of Being Human*. Blackwell, 1988.
Hocking, William Ernest. *The Meaning of Immortality in Human Experience*. 1937. Rev. ed. Greenwood Press, 1957.
Hunt, Peter. *An Introduction to Children's Literature*. Oxford University Press, 1994.
Kahn, Peter H., and Stephen R. Kellert, eds. *Children and Nature: Psychological, Sociocultural, and Evolutionary Investigations*. MIT Press, 2002.
Kellert, Stephen R., and E. O. Wilson, eds. *The Biophilia Hypothesis*. Island Press, 1995.
Kidd, Kenneth B. "A is for Auschwitz: Psychoanalysis, Trauma Theory, and the 'Children's Literature of Atrocity,'" *Children's Literature* 33, no. 1 (January 2005): 120–49.
Kidd, Kenneth B. *Freud in Oz: At the Intersections of Psychoanalysis and Children's Literature*. University of Minnesota Press, 2011.
Love, Glen A. *Practical Ecocriticism: Literature, Biology, and the Environment*. University of Virginia Press, 2003.
Major, William, and Andrew McMurry. "Response of William Major and Andrew McMurry." In *Teaching Environments: Ecocritical Encounters*, edited by Roman Bartosch and Sieglinde Grimm, 227–234. Peter Lang, 2014.
Marx, Karl. *Economic and Philosophic Manuscripts of 1844*. https://www.marxists.org.
Mazel, David. "Ecocriticism as Praxis." In *Teaching North American Environmental Literature*, edited by Laird Christensen, Mark C. Long, and Fred Waage. MLA, 2008.
Morton, Timothy. *Ecology without Nature: Rethinking Environmental Aesthetics*. Harvard University Press, 2007.
Myers, Olin Eugene, Jr. (Gene), and Carol D. Saunders. "Animals as Links toward Developing Relationships with the Natural World." In *Children and Nature: Psychological, Sociocultural, and Evolutionary Investigations*, edited by Peter H. Kahn and Stephen R. Kellert. MIT Press, 2002.
Myers, O. Gene. *The Significance of Children and Animals: Social Development and Our Connection to Other Species*. 1997. Rev. ed. Purdue University Press, 2007.
Nicols, Ashton. *Beyond Romantic Ecocriticism: Toward Urbanatural Roosting*. New York: Palgrave Macmillan, 2011.
Nodelman, Perry. *The Hidden Adult: Defining Children's Literature*. Johns Hopkins University Press, 2008.
Oppermann, Serpil. "Ecocriticism's Phobic Relations with Theory." *ISLE* 17, no. 4 (Autumn 2010): 768–70.
Parham, John. "The Poverty of Ecocritical Theory: E. P. Thompson and the British Perspective." *New Formations: A Journal of Culture, Theory and Politics* 64 (2008): 25–36.
Plumwood, Val. *Feminism and the Mastery of Nature*. Routledge, 1993.
Reynolds, Kimberley. *Radical Children's Literature: Future Vision and Aesthetic Transformations in Juvenile Fiction*. Palgrave Macmillan, 2007.
Richerson, Peter J., and Robert Boyd. "Culture is Part of Human Biology: Why the Superorganic Concept Serves the Human Sciences Badly." In *Science Studies: Probing the Dynamics of Scientific Knowledge*, edited by Sabine Maasen and Matthias Winterhager. Biefeld: Transcript Verlag, 2001. Draft 3 (7.17.01).
Rigby, Kate, and Axel Goodbody. "Introduction." In *Ecocritical Theory: New European Approaches*, edited by Axel Goodbody. University of Virginia Press, 2011.

Roback, Diane, ed. "All-Time Bestselling Children's Books." *Publisher's Weekly* 248, no. 51 (December 17, 2001). Accessed May 16, 2014. https://www.publishersweekly.com/pw/by-topic/childrens/childrens-industry-news/article/28595-all-time-bestselling-children-s-books.html.

Rose, Jacqueline. *The Case of Peter Pan, or, the Impossibility of Children's Fiction.* Macmillan, 1984.

Rosenberg, Alexander. *Sociobiology and the Preemption of Social Science.* Johns Hopkins University Press, 1980.

Roszak, Theodore. "Against Single Vision." In *The Green Studies Reader: From Romanticism to Ecocriticism*, edited by Laurence Coupe. Routledge, 2000.

Soper, Kate. *What Is Nature? Culture, Politics, and the Non-Human.* Blackwell Press, 1995.

Swanigan, Pamela. "The Case Against Agenda." In *Teaching Environments: Ecocritical Encounters*, edited by Roman Bartosch and Sieglinde Grimm. Frankfurt: Peter Lang, 2014.

Tugwell, Simon. *Human Immortality and the Redemption of Death.* Templegate, 1990.

White, E. B. *Letters of E. B. White.* 1976. Updated Ed. Dorothy Lobrano Guth, ed. HarperCollins, 2006.

White, Lynn, Jr. "The Historical Roots of Our Ecologic Crisis." In *The Ecocriticism Reader: Landmarks in Literary Ecology*, edited by Cheryll Glotfelty and Harold Fromm, 3–14. Athens: University of Georgia Press, 1996.

Wilson, E. O. "For Sociobiology." *New York Review of Books*, December 11, 1975. http://www.nybooks.com/articles/archives/1975/nov/13/against-sociobiology.

Wilson, E. O. *Biophilia.* Harvard University Press, 1984.

Wilson, E. O. "What Is Sociobiology?" *Society* 15, no. 6 (September–October 1978): 10–14.

Wordsworth, William. "Ode: Intimations of Immortality." In *Major British Poets of the Romantic Period*, edited by William Heath. Macmillan, 1975.

Yoke, Carl B. "Why Was I Born If It Wasn't Forever?" In *Death and the Serpent: Immortality in Science Fiction and Fantasy*, edited by Carl B. Yoke and Donald M. Hassler. Greenwood Press, 1985.

1 Pan, pantheism, and natural immanence

The beginning of children's-literature syncretism in *The Wind in the Willows* and *The Secret Garden*

In starting my ecocritical examination of immortality in children's fantasy with Kenneth Grahame's *The Wind in the Willows* (1908) and Frances Hodgson Burnett's *The Secret Garden* (1911), I begin at what many people would consider the beginning.

Grahame's classic is an episodic, sometimes picaresque book about the leisurely riverside adventures of a water rat, a mole, and a toad, while Burnett's novel features two sickly, fretful 10-year-olds who achieve health and happiness by running around outside and digging in the dirt. These two giants of the genre not only epitomize children's literature as it is generally thought of, but also, through their unassailable canonicity and commercial success, have acted as crucial precedents and permission-givers. Children's literature of this era marks the start of the genre as it has come to be known in the popular mind: full of the delight of the outdoors, of children interacting with animals, of animals that are at once human and their own species in such a way as to erase any meaningful distinction between them, of character arcs and moralities shaped by these surroundings and interactions.

Critics are nearly unanimous in seeing the late-Victorian and Edwardian era as the period in which, as Gillian Avery notes, "the Romantic image of the uncorrupted child" was "generally admitted into children's literature"[1] and in which children's authors "wholeheartedly embrac[ed] the 'Child of Nature' paradigm."[2] David Sandner comments that the literature of this era "display[s] many hallmarks of its Romantic heritage," including "the visionary, pantheism, the wise child, the imagination unbound."[3] Humphrey Carpenter has famously termed the years 1860–1930 a "Golden Age" for children's literature, one marked by a "theme of ... Arcadia, the Enchanted Place, the Never Never Land, the Secret Garden," by often being "set in a distant era when things were better than they are now,"[4] and by "the possibilities ... of Escape."[5] Sarah Gilead believes that it is at this point that "what is at the heart of 'canonical works of children's literature' formed as a 'dynamic tension between idyllic, childhood realm, and

DOI: 10.4324/9781003784210-2

the adult world that subverts it.'"[6] Seth Lerer goes so far as to declare that the pastoral, sunlit portrayals in the children's literature of this era have become "our default mode of childhood."[7]

Notwithstanding Marah Gubar's persuasive account of how "children's authors ... were censured in their own time for *failing* to promote a Romantic ideal of primitive simplicity,"[8] it is hard to deny that by the early-20th-century publication of *The Secret Garden* and *The Wind in the Willows*, "the rude brow of Rousseau," which had "lour[ed] on the horizon" of John Locke's more rationalist terrain,[9] had materialized into the genre's full incorporation of that author—and, more, the dominance of Rousseauean precepts about how children should spend their time and best be educated. Indeed, many books for young people from this point on resemble nothing so much as literarized versions of Rousseau's fictional tutee Emile, who, "brought up in the country ... run[ning] around tree-climbing and leaping brooks"[10] and having been physically, psychologically and morally "strengthened by ... constant exercise under the guidance of nature herself," learns "to use the natural tools which are within our reach and adapted to our senses."[11]

One might assume, from all this, that nature would have been a primary focus for children's literature critics over the past 100 years, especially the many who analyze Golden Age literature such as *The Secret Garden* and *The Wind in the Willows*. One would expect to see the field of ecocritical examination, by that name or any other, trammeled into a bog.

In fact, as I observe in my introduction, it has not been. Almost without exception, critics have followed the practice at the heart of Lawrence Buell's perplexed 600-page complaint *The Environmental Imagination*, of acting as though "in 'good' writing ... biota has only a bit part" and "the literary naturescape exists for its formal or symbolic or ideological properties rather than as a place of literal reference."[12] As the faint pejoration of Carpenter's capitalized "Escape" forebodes, the phase of North American and British literary studies when children's literature became a legitimate subject was not one that found academics in a mood to accept kindly the Romantic sensibility or its generic offspring. Most critics mention the "child of nature" paradigm only to disparage it or mine it for problematic constructions of childhood. In Jane Darcy's apt summation,

> [C]riticism often focuses on what is perceived as a general tendency on the part of children's writers in this period to idealize and romanticize childhood by depicting children or childlike figures leading an idyllic life in a rural environment. Because such representations do not appear to reflect the realities of emerging democracy in what had become a complex industrial and urban society, they are seen as backward looking and as evading the "truth."[13]

Nature, meanwhile, is cast as a vehicle for that contemptible feeling, nostalgia. It is hardly possible to find a critical examination of these two books (or of late-Victorian and Edwardian children's literature in general) that does not feature the word "nostalgia," almost always used disdainfully. As Darcy notes, the fact that "both texts ... in their different ways derive much from a Romantic tradition in their representation of childhood and nature" has been used to "impl[y] nostalgia, sentimentality—an absence of real feeling—and, more importantly, a convolutedness and insincerity in the texture of the writing."[14] Miles Kimball represents this view when he writes, "The nostalgia for childhood in works influenced by the Romantic child approaches a psychological condition—nostalgia, in other words, in its literal sense as a sickness like neuralgia—and this nervous condition has less to do with childhood than with adult attitudes toward it, both on the part of adult readers and adult writers."[15]

The critical reading-over of nature, which seems proportionately more remiss as the overtly ecocentric content of a text or genre increases and therefore veritably delinquent in the case of children's literature, appears outright egregious when applied to texts as forcibly nature-themed as *The Secret Garden* and *The Wind in the Willows*. They are, after all, called *The Secret Garden* and *The Wind in the Willows*. However, among the 43 specific critical analyses and numerous theoretical approaches that editors Jackie C. Horne and Joe Sutliff Sanders list in their introduction to *Frances Hodgson Burnett's* The Secret Garden*: A Children's Classic at 100*, the sole mention of nature is relegated almost to a footnote at the end of the Critical Approaches section: "A handful of articles address issues other than adaptations of the novel's construction of femininity, imperialism, or sexuality," observe the editors, "[including] Jane Darcy's analysis of the depiction of nature."[16]

The Wind in the Willows fares just as poorly. When Horne and Donna White, editing *Kenneth Grahame's* The Wind in the Willows*: A Children's Classic at 100*, suggest as one of the "avenues waiting to be explored" whether "the pantheism of *Wind* [can] help explain the less overtly religious sublimity of later children's fantasy,"[17] we might take this to mean that more basic ecocritical examinations have already been carried out. This is not the case. Horne and White mention critical trends from biographical criticism in the early going through the obligatory psychoanalytic phase in the 1990s, with class, gender, Englishness, and traditional close readings in between. But of the 45 critics they list by name, for analyses of nature they can cite only "Jane Darcy, who sees in Grahame's depiction of nature in *Wind* the potential for inspiring contemporary readers to recognize their need for connection to the green world."[18] This refers to the same lone Jane Darcy essay mentioned in *The Secret Garden*'s centenary.

One would expect that Grahame's numerous and overt allusions to Romantic poets would encourage scholars to focus on the natural content of his book. Instead, these treatments generally skirt the Romantics' identities as nature poets and work with other elements of their philosophy, as in Lesley Willis's essay "A Sadder and Wiser Rat / He Rose the Morrow Morn," which "suggests that Grahame's allusions to Romantic poems highlights the novel's 'central tension' between 'the paramount value of home and community and the merit of accepting … one's own limitations' and the 'desire for adventure and change.'"[19]

The centenary compilations correct some of this deficiency. The Grahame book offers essays on the presence of Pan by David Rudd and a contradistinction between Techne and technology by Deborah Dysart-Gale, while the Burnett book offers Alun Morgan's valuable exploration of the secret garden as a place of transformation. Nonetheless, these eco-critical approaches continue to take a back seat to more recondite analyses such as those of "the construction of heteronormative masculinity [and] the appeal of this very English novel [*The Wind in the Willows*] to Chinese readers."[20]

I agree as much as anyone that *The Secret Garden* and *The Wind in the Willows* typify the mainstream of modern canonical children's literature. I agree, too, that a great deal of what they contain and typify is a "Romantic" conception of the "child of nature"—though obviously I view this in a different and more positive light, and I also see this conception as more nuanced and complex than the Romantic writers depicted or is commonly perceived. Indeed, these two premises constitute a large part of my thesis and this chapter. What I would say, however, is that so long as the "nature" side of this formulation goes almost wholly unexamined, the idea itself cannot possibly have any real meaning. And if, as critical consensus holds, this idea sits at the heart of children's literature from the late Victorian era onward, then the genre cannot be said to be properly understood.

So what conception of the child in relation to nature do these two books convey, foster, and reinforce? It is, above all, a biotheistic one. As Darcy notes, the two novels are "alike in celebrating a life close to nature and in investing the natural world with moral significance and with a quasi-religious mysticism."[21] I would go beyond the "quasi" and say that reverence for nature, and even formalized nature-worship, forms a dominant mode in these books. *The Secret Garden*'s main characters, Mary and Colin, develop a ritual of chanting to attune themselves to "the Magic," out of which, as Colin earnestly declares, everything is made, from trees and flowers to birds and badgers—and, signally, people. The Magic makes the sun itself rise and creates in humans "a strange feeling of being happy."[22]

Meanwhile, *The Wind in the Willows*' central character Mole, as David Rudd convincingly argues, is entranced by the nature god Pan from the very first chapter, when he leaves his domestic duties to go live on the River Bank, bends his ears to the reeds, and catches "something of what the wind went whispering so constantly among them."[23] Mole and his friend Rat later encounter this god and, "crouching to the earth, [bow] their heads and [do] worship."[24]

These "pagan" modes of worship—which term I shall use as an umbrella for the various pre- and non-Christian beliefs depicted in the two books—echo the biotheist Albert Schweitzer's proposal that "life in its creative aspect constitutes a sacred force."[25] The biotheism in these books and in children's literature in general is, moreover, more realistic and thus more courageous than Schweitzer's, in that it doesn't, as Michael Martin claims of Schweitzer, "avoid recognizing that killing by non-human animals and plants is integral to life in its sacredness."[26]

It is also a syncretic belief system, one that was quickly to become at once endemic and peculiar to children's literature. This is unsurprising, in both the general and the specific. To begin with, deism and alternative spiritualisms pervaded *fin de siècle* British society: as Humphrey Carpenter and others have pointed out, "almost without exception, the authors of the outstanding children's books that appeared between 1860 and 1930 rejected, or had doubts about, conventional religious teaching."[27] Further, both Burnett and Kenneth Grahame openly explored non-Christian belief systems: Burnett theosophy and Christian Science,[28] among others; Grahame the classical mythology that infuses his *Pagan Papers* (1893).

Burnett, who as a child absorbed ideas of animistic theism by reading James Fenimore Cooper's Leatherstocking Tales,[29] presumably recovered enough from her disappointment over the less glamorous reality of Native Americans in her adoptive Tennessee[30] to transmit values and perceptions of nature in *The Secret Garden* that owe as much to some North and South American First Nations' nature-oriented paradigms of sacredness as to the Indian fakirism that Mary and Colin cite. The ideas of pre-Christian gods and precepts that permeate *The Wind in the Willows* descend through the Romantic era from Europe's strongest syncretic traditions and through Romanticism. As Albert J. Kuhn observes,

> Especially in England … there was a zealous desire to resurrect divinities long forgotten and to find in them and their exploits a relevance for the modern age … For the most part, these theorizers on myth … were syncretists. They sought … to show that beneath the seemingly disparate and heterogeneous elements of ancient universal mythico-religious and historical traditions there lay a harmonious tradition.[31]

As this syncretic tradition expresses itself in *The Secret Garden* and *The Wind in the Willows*, its uniting force is the representation of a worldview and human psychology as evolution shaped it: alert to life; fascinated by a type and level of complexity found only in the biotic and abiotic natural community; self-perceived as an equal, interactive, interdependent member of that community; and capable of full gratification and self-realization only through direct and continual relationship with that community.

This is not, needless to say, the belief system generally ascribed to modern Western society, much less to the literature that society gives its young people. It is heretical not only to the exceptionalist doctrines of Judeo-Christianity but to its secular twin, humanism; to the techno-industrial capitalism and urbanization to which Western Christianity and humanism gave rise; to the theories of Freudian psychology and most non-Nietzschean philosophies; and even to what E. O. Wilson calls "science rampant resisted as 'scientism', an unpleasant doctrine."[32] One can see within it the germ of Counter-Enlightenment thinking, preserved and transmitted from the 1700s to present day. By any measure, it is ostensibly both anomalous and anachronistic. Yet in different forms and to varying degrees, this belief system is a mainstay of the literature that Western society has been giving to the young of the species for more than 100 years now. It is also, in Alun Morgan's words, "prescient in terms of the support for these very ideals now emerging from the relatively new fields of environmental and conservation psychology … positive psychology … and the applied field of environmental education."[33]

Critical to the type and degree of the syncretic biotheism in *The Secret Garden* and *The Wind in the Willows* are the daringly overt non-Christian conceptions of immortality to be found in both books. These separate them out from earlier texts of natural theology such as Johanna Spyri's *Heidi* (1880) and pave the way for representations of biophilia in children's literature that reach beyond a mere Rousseauean prescription for an active, outdoorsy childhood and into the realm of a truly alternative worldview, complete with a spiritual or religious paradigm that restores humanity to a position of reverent ecological parity within the natural matrix.

The syncretic aspect and alternative conception of immortality appear differently in the two books. They are both more explicit and more contained in *The Wind in the Willows*, emerging from a subtle intimation through scattered allusions to Romantic poets into open neoclassical nature-worship in the "Pan" chapter, "The Piper at the Gates of Dawn." In *The Secret Garden*—which, by virtue of its more extensive focus on immortality, will receive the preponderance of my analysis—Burnett writes, in Fred Inglis's phrasing, "straight from the well-springs of Romanticism

… [She] reinvents a pagan Garden of Eden for the children where culture is detached from labour and returned to creativeness."[34]

While Holly Blackford is undoubtedly right in seeing a strong presence of the Eleusinian Mysteries in *The Secret Garden*, and in reading "the Greek myth of Persephone and Demeter [as] infus[ing] every step of Mary Lennox's journey,"[35] the more explicit paganism, as with *The Wind in the Willows*, centers around the book's Pan figure, the peasant boy Dickon. This "Pan-boy," as Inglis terms him, "tends and understands in the name of mystery that which Romanticism sought to keep intact from science."[36] Burnett's late-career view that "spiritual relationships are possible between the dead and the living, and that such relationships enhance rather than detract from life"[37] ultimately extends through all of her main characters into ritualized nature-worship, as narrative expressions of pantheism and biological coinherence, and a view of death and afterlife that merges the reincarnative and transmutational with the more literal modes of biological dormancy and renewal. John Rowe Townsend notes,

> There is something about *The Secret Garden* that has a powerful effect on children's imaginations: something to do with their instinctive feel for things that grow, something to do with their longing for real, important, adult-level achievement. … Perhaps it required someone like Mrs. Hodgson Burnett, who was no simple Christian (she dabbled in various unorthodoxies, and *The Secret Garden* contains clear indications of belief in some kind of Life Force) to bring forward this new, significant and potentially subversive doctrine.[38]

As the book reaches its most forceful explication of thesis in the final chapters, "[t]he awakening of an inner Dickon and therefore pantheism sets the return of spring in motion."[39] It gives voice to Burnett's view of gardens as a site of essentialist recycling, the pantheistic idea that "we are made one with what we touch and see," as expressed in her essay "In the Garden" (also the title of the final chapter of *The Secret Garden*): "As long as you have a garden you have a future, and as long as you have a future you are alive."[40]

Inasmuch as these two books are precedents as well as representatives, I see as an important part of my task to extract and recontextualize the biophilic elements that we have come to take for granted in children's literature. This will shift my focus in the early going from analyzing immortality to conducting a wider historical re-examination—one that includes, to some degree, performing an ecocritical counter-reading of textual elements that may connect only indirectly to immortality.

Saving the children, biophilically

Critics are generally agreed that the shift within children's literature represented by *The Wind in the Willows* and *The Secret Garden* signifies the consolidation of a "new normal." They note that the absence of overt didactic purpose and presence of imaginative gratification in books such as Grahame and Burnett's, along with *Alice in Wonderland* and *Peter Pan*, represents a significant change from the preceding 150 years of texts given to children to read. As Humphrey Carpenter expounds,

> The greater number of children's books published in England between the 1740s and the 1820s were sternly moral, using simple stories to convey whatever ethical message was then in fashion, whether it be that hard work always leads to improvement in one's financial and social position (a moral much favoured from the 1750s to the 1780s), or that idle and thoughtless children would soon die an unpleasant death and then suffer everlasting torment in Hell (a message that became fashionable with the rise of the evangelical movement early in the nineteenth century).[41]

Those who have sought to explain why there was a delay of 100 years or so between the Romantic precepts of the early 1800s and their emergence in children's literature in the early 1900s usually cite the late-Victorian and Edwardian cultural-ontological instability wrought by Industrialism and Darwinism. Deborah Dysart-Gale observes that "The world in which Grahame and his audience lived was a world in transition, one in which technology and modernity disrupted lives and displaced characters"[42] and in which "familiar, rural lifestyles appeared to be fading under the new social and political organization of modernity, and misgivings about the future were confirmed with the beginning of the Great War."[43]

Books such as Grahame's, Dysart-Gale posits, were popular because they "succeeded in addressing [their] audience's concerns regarding technologically driven social change."[44] Humphrey Carpenter, likewise, notes that modern-day children's literature was born at the same time as "[t]he realization … that the apparent triumphs of the Industrial Revolution had produced widespread misery for the working classes."[45] Alun Morgan echoes this, characterizing the late-Victorian and Edwardian eras as a time when the social and environmental devastation of the Industrial Revolution was making itself clear, resulting in leading cultural figures such as the pre-Raphaelites "start[ing] to preach that the route back to a healthy society lay through a flight from industrialization, a return to medievalism, and the revival of old methods of craft."[46]

Despite this consensus, however, critics have gone on to draw oddly vague and aleatory conclusions about the fact that "the work of the great children's writers… formed some sort of discernible pattern of ideas and themes."[47] Perhaps it is because, as Greg Garrard points out, Romanticism had been traditionally defanged by the historicists and deconstructive formalists—for whom "the phrase [radical pastoral] had come to appear as oxymoronic as 'military intelligence' or 'humane slaughter'"[48]—and therefore the Romantic inheritance of children's literature has usually been viewed as a "taming" element, that critics have failed to situate children's literature of this period within the broader "green politics" that Peter Gould observes as arising in the late Victorian period. Echoing the children's literature critics, Gould cites this politics as being born of "a response to economic depression and social unrest centred around a critical examination of, in particular, the philosophy of industrialism, the relations of the individual to their social and physical environment, and the function of the city."[49]

The relegation of this response to "nostalgia" is part of a persistent but wrongheaded social pressure to invalidate genuine, evolution-based mourning, not to say panic, over the loss of contact with natural surroundings. Jane Darcy reminds us that "in approaching the representation of nature in children's books of the late Victorian and Edwardian period … it has an immediate and real as well as a mythical element."[50] She cites Marxist scholar Raymond Williams' comment that "despite the tendency of each generation to site the mythical lost Golden Age in the previous generation, there were also those who had 'lived' that history, and to whom any change was very real."[51] She cites also Williams' important datum that "[a]t the beginning of the century, a third of all workers were employed in agriculture; in mid-century, a fifth; by the end of the century, less than a tenth."[52] In Peter Gould's view, this deracination caused "[e]arlier 'back to nature' elements … [to be] transformed from the 1880s into a radical, populist political agenda."[53]

If we consider children's literature as eligible for membership in this agenda, it appears plausible, or even evident, that children's literature at this time became a site of resistance to the rupture from the biotic community in which we evolved over the millennia. In the domain of cultural arts, we could even argue that it became the *main* site of resistance. So from an ecocritical perspective, it seems clear that what *The Wind in the Willows* and *The Secret Garden* initiated, and what has been in large part responsible for their enormous and enduring success, was their role as key cultural transmitters of biophilia—as receptacles in which the newly alienated population could "hide" its increasingly disjunctive "affinity for life and lifelike processes."[54]

Looking from this ecocritical perspective to the novels' rhetorical function in "persuading children to accept certain premises about the world and teaching them how to live and what to value within it,"[55] it behooves us to ask what, exactly, Kenneth Grahame and Frances Hodgson Burnett teach through *The Wind in the Willows* and *The Secret Garden*.

The Wind in the Willows starts when the main character, Mole, abandons spring-cleaning for a life of excursions and adventures with Rat on the riverbank. He is introduced with a high degree of anthropomorphism, working in a human-like way to clean his house to the height of urban British housekeeping standards. This includes not only sweeping and dusting, but also whitewashing—the method representing the most complete rejection of dirt, and therefore of nature, and especially unnatural for a character who literally lives underground. As he works away, with "aching back and weary arms," his animal nature emerges and then overcomes him. "Spring … with its spirit of divine discontent and longing"—calls to him so "imperiously" that he throws down his cleaning implements, says "'Bother!' and 'O blow!' and also 'Hang spring-cleaning!' and bolt[s] out of the house without even waiting to put on his coat."[56] In other words, the "inciting incident" of the book is no more than an overwhelming compulsion to go outside. Mole spends the rest of the book excursing with Rat in a boat and having various outdoor adventures.

In *The Secret Garden*, sickly, sour Mary Lennox becomes strong and vital by being forced to play outdoors and then tending to the locked garden to which she finds the key. After she has become as "wick" (healthy) as the plants and her animal-charmer friend Dickon, she helps her sickly, sour first cousin Colin become similarly strong and vital by going outdoors and tending to the garden as well. Fundamentally, nothing else happens in either book. It is worth keeping this in mind when considering that these two books are the templates and permission-givers for modern children's literature, as well as perennial bestsellers.

Granted, numerous critics have found a conventional moral purpose in the two novels. For instance, Kathryn V. Graham says of *The Wind in the Willows* that "[t]he novel's episodes promote friendship, courtesy, competence, courage, and generosity in an idyllic world where sex, work, violence, and death are beyond the horizon."[57] (This is incorrect: the last three of these, at least, are easily found in most chapters of the book. Sometimes they appear quite graphically, as in the frightening scenes in the woods near Badger's house and the gory attack of the stoats near the end of the book.) Jackie C. Horne and Joe Sutliff Sanders declare confidently that the overt message of *The Secret Garden* is that "Mary must learn to empathize with others."[58] Gretchen Holbrook Gerzina, similarly, reads many passages with a Christian bent in her annotation of *The Secret Garden*, saying, for example, that "In naming the housemaid 'Martha' and the newcomer 'Mary,' Burnett draws on the New Testament … Mary … was considered a true believer … Dickon acts as a healer, who helps to raise up Colin, like Lazarus, from a state of expected death."[59] Her annotations emphasize

Mary's "healing" as equated with her "learn[ing] to help others."[60] Others, as noted in the introduction to this chapter, have read feminist, imperialist, classist, and Freudian meanings into both novels.

Such readings, however, seem more than a little Procrustean. Neither book features a character arc as such: Rat and Mole are morally and characterologically unchanged—as is the unregenerate Toad—at the end of *The Wind in the Willows*. Although Mary and Colin have changed substantially by the end of *The Secret Garden*, the change is to their emotional and physical states, not their characters: both children remain strong-willed, impatient, imperious, and relatively self-centered. Even after Mary's revivifying discovery of the walled secret garden, deep into the book, in the defiantly named Chapter 16, "'I won't!' said Mary," she is acerbically described as "not a self-sacrificing person."[61] Nor does she become one by the end: she remains "angry, unsympathetic little girl,"[62] one whose blunt, direct, unvirtuous ways prove the best medicine for rejuvenating her sickly cousin Colin. She receives authorial approval and amusement for her technique of outshouting Colin; her truth-telling slows his slide toward self-willed invalidity and separation from the natural world.

In sum, Mole and Rat's and Mary and Colin's behavior accords very little with the orthodox messages of moral and social conformism that these critics would seem to impose. As John Rowe Townsend remarks, "the virtues that Mary and Colin painfully attain ... are not Victorian virtues. The Victorian ideal was that children should be good and do as they were told ... it was no part of this ideal that children should be self-reliant, constructive, inner-directed."[63]

As with other connections between cultural upheaval in British society and the biophilic turn of children's literature, the simplest and most obvious interpretation would seem to be a face-value ecocritical one—namely, that Grahame and Burnett are participating in a larger social effort to get children outdoors. Alun Morgan observes,

> The late Victorian and early Edwardian periods saw the emergence of the "child saving movement" in response to the severe degradations wrought by the industrial revolutions and massive urbanization very much in evidence at that time... . [T]his movement was very much focused on promoting children's healthy development and remediation through play and self-expression, particularly in natural settings.[64]

Seen as "resonan[t] with contemporary thinking in psychology and education" in our own disrupted era of "post-Industrialization, peak oil crises, global climate change and ... nature deficit disorder,"[65] the two books appear to be delivering a simple set of instructions to the child reader. Go run around outside. Dig in the dirt. Look at the sky. Listen to the birds.

Look at the plants. Breathe. Sweat. Eat a lot. Play a lot. Sleep deeply. Revere the biotic and abiotic natural world, yourself included, and understand that it and you belong to an immanence that bears little resemblance to—in fact, contradicts—the morality-based, abstract Heavenly immortality that Christianity offers.

Although this depiction of ideal childhood (inasmuch as the clearly adult Rat and Mole nonetheless act the role, or at least fill the role, of children) unquestionably owes much to Jean-Jacques Rousseau's exhortations in *Émile* to "[t]each the child to live rather than to avoid death"[66] and to "[s]end your children to renew themselves ... send them to regain in the open fields the strength lost in the foul air of our crowded cities,"[67] the rebelliousness of these protagonists' passionate connection with the outdoor natural environment speaks to how political and transgressive the character set is that attends the human-as-animal worldview and the "health through fresh air" prescription. Willfulness, formerly a trait which children's authors vigorously discouraged, is now valued as a form of determination. Industriousness is recast as gratifying, play-like interactions with the earth and its denizens. Abstemiousness is derided and appetite applauded; abstract philosophizing is depicted by omission in Grahame's book as unnecessary and by commission in Burnett's book as dangerous. And any inclination to religious sentiment is firmly redirected to reverence for nature, with both books openly dismissive of monotheistic precepts.

Inherent within and attendant upon this set of instructions is a rejection of Western allopathic medicine as a route to health. As Morgan says of *The Secret Garden*, the books invoke a "self-conscious awareness of the 'outer realm' of the biophysical environment ... as an agentic character in its own right, one that acts therapeutically upon the chief protagonists."[68] Here, as with the regimen of outdoor play and exercise, we can see an early inoculation of the child reader—and their descendants—against the "unnatural" routes to life extension that would be increasingly propounded by immortalists starting in the 1960s and promoted ever more stridently at the turn of the millennium by the commercial arms of the genetics and pharmaceutical industries.

The Wind in the Willows proposes no afterlife beyond its inclusion of the Greek deity Pan. Grahame settles for portrayals of vital animal health and activity in *this* life, rooted to the earth. Burnett, after her son Lionel's death, rejected allopathy and pursued more Rousseauean modes of medicine, to the extent that many critics and biographers regarded her as a Christian Scientist. (It was an appellation that Burnett rejected only circumspectly; see Cadwallader 121, Gerzina 240.) She also, as Morgan notes, actively participated in "child-saving" work, such as visiting the Invalid Children's Aid Association and giving "a specially designed carriage to a young girl with a spinal deformity so she could be wheeled outdoors."[69]

The Secret Garden scorns the "condition-curing" approach of allopathic medicine: Mary's best cure, for everything from sickliness to plainness to fretfulness to loneliness, is to go outside and play. With this cure undertaken, Mary soon becomes fatter and stronger. She notes, "I used always to be so tired. When I dig I'm not tired at all.'"[70] Dickon concurs, telling her that the exertion is good for her. He himself never catches a cold, not having been brought up "nesh" enough; his mother, he tells Mary, attributes this to his habit of living almost entirely outdoors and "sniffing up" fresh air. As the days pass, Mary instinctively follows the prescription of fresh air and outdoor living.

Further, observing the wealth of positive changes in herself, Mary mentally contravenes Dr Craven's strictures regarding her lifeless cousin Colin and thinks that if the garden and outdoor play have helped her, perhaps they will also be good for Colin. In Colin, Mrs Sowerby diagnoses what would in this day and age be called extreme nature-deficit syndrome: "[She] says there's no reason why any child should live that gets no fresh air an' doesn't do nothin' but lie on his back an' read picture-books an' take medicine."[71] Colin's doctor can keep him only in a state of suspended animation (minus the animation) with his medications, braces, and rest-cures, while the cure of communion with nature takes Colin, like Dickon, to an exalted state of felt immortality and union.

By embracing biophilic principles and rejecting the nature-segregative social-economic-religious beliefs of the Enlightenment period, Grahame and Burnett institute in children's literature the acceptance of—and eventually the expectation of—a turning-away from Christian morality and hortatory promotion of behavioral conformity. In point of fact, both Burnett and Grahame would appear to be trying to "save the children" from many of the outcomes that much children's literature up to that point had striven to achieve. These include a blind subscription to Abrahamic assumptions of human superiority and separation from the rest of the created world (alloyed but probably not ameliorated in its effects by the self-loathing tenet of Original Sin); the religious-psychological construct lying (also uneasily) beside a hardening and straitening understanding of scientific precepts; and the rapid, enormous advances in technology which, as Lynn White, Jr., puts it, "[f]rom the eleventh century onward...has (in) Occidental culture...increased in a steady crescendo."[72] The conjoining of science and technology, the surface reconciliation of Darwinism and Christianity, the supposedly eager embrace of industrial capitalism, and other givens of today's social constructions were new and fluid when books such as Burnett's and Grahame's interceded, as I propose they did, with their resistant, subversively conservative biocentric paradigms.

Thus, it was arguably these two books that first and most firmly instilled in children's literature the animalistic, evolutionary conception of the human species and its life cycle. This conception inseminated in the genre the strong resistance to both the scientific, pre-death immortality and Abrahamic, post-death immortality that I find and examine in later chapters.

Edwardian ecology: Romanticism with teeth

In both *The Secret Garden* and *The Wind in the Willows*, the most reified representation of biotheistic syncretism and alternative immortality comes in the presence of Pan, the Greek god of pastures, fertility, the wild, shepherds, and music. Pan *qua* Pan appears in *The Wind in the Willows*, as the demi-god who is found watching over Little Portly, the lost young otter, and to whom Rat, Mole, and all the inhabitants of nature pay worship. In *The Secret Garden*, he appears as Dickon, the peasant boy who lives on the moor, rescues orphaned animals and charms them with his pipes, and is adept at working the nature-revivifying "Magic" in which Mary and Colin eventually situate eternal life.

David Rudd remarks on Pan's unique capacity for "blending" as a "hybrid being that melds the natural (in both the bestial, and unspoilt, senses) and the supernatural (spiritual, disembodied)"[73] Patricia Merivale, similarly, comments on the fact that the etymological conflation of the Greek *pa-on* (grazer), with "*pan*" (everything), "made Pan not only merely the god of Arcadia, but of the whole of nature … not merely the god of shepherds, but, in Christian myth, either the devil, as might be expected … Christ, as shepherd-god himself."[74]

On the face of it, Pan's presence in these two children's books seems perhaps unworthy of comment, since, as Rudd points out, "In the fin de siècle he [Pan] certainly was everywhere."[75] Miles Kimball and William Greenslade concur. Kimball writes, "The popularity of the Greek god Pan as a representative of Aestheticism or Bohemianism was well established in the decades approaching the twentieth century,"[76] while Greenslade notes, "Undoubtedly, at the fin de siècle, the literary 'public' *had* heard a great deal of 'Pan'."[77] Pan's omnipresence makes it easy to gloss over the significance of his appearance in a book for children, and to conclude, as *Wind in the Willows* annotator Seth Lerer does, that "It is no accident that the embossed cover of the 1908 edition of *The Wind in the Willows* should have Pan as its icon."[78]

However, Pan's presence and function in *children's* books warrants more careful consideration. While, as Beverly Lyon Clark notes, "[c]hildren and childhood were less segregated from adults and adulthood in the nineteenth century,"[79] they had begun to separate by the turn of the century. By the publication of *The Wind in the Willows* and *The Secret Garden*, children's literature was firmly set as a category hedged about by adult protections. For instance, critical readers of Burnett's later short stories questioned "the appropriateness of death as a subject matter in children's literature," and scholar Madelon S. Gohlke writes of *The Secret Garden* that "A book so obviously preoccupied with death … may seem an odd candidate for children's literature."[80]

In short, it was far from inevitable that a pagan element such as Pan would follow from adult *fin de siècle* culture into the protected domain of children's literature. Miles Kimball believes that "Pan shows up in children's literature as a simultaneously wistful and dangerous figure" who demonstrates a metaphorical and mythical transgressivity"[81] by "going against the grain of modern society by escaping from it."[82] My view accords more with that of Greenslade, who sees the "pagan renovation" as an attempt "to out-manoeuvre Christian teaching."[83] The fact that one of the main strains of Victorian culture to find its way to the heart of children's literature was paganism and pantheism, in the new context of Darwinism and an increasing acceptance of human-as-animal that threatened the Christian and Industrialist status quo, speaks to the urge toward biophilia in the genre.

Such uses of the Counter-Enlightenment's neo-pagan strain, with their resonances of "The world is too much with us," are generally held to confirm the "Romantic inheritance" of Golden Age children's literature. This, in turn, is generally used as a basis for consigning the genre to the ranks of escapism—in David Sandner's words, a literature that "presented itself … as without purpose except delight."[84] However, if we are to note the books' considerable debt to the Romantics, we must understand the attendant implications in the light of two ecocritical insights.

The first of these is the ecocritics' (mostly signaling Jonathan Bate's) recasting of Romantic texts, whose radical component function academics historically have dismissed, as being ecologically and theistically revolutionary: "congenial surroundings from which to launch a coherent attack on the critical orthodoxies of today."[85] If we are going to accept the resuscitated subversive powers of Romantic paganism, we must also rehabilitate the "conservative" reputation of children's literature, something that has yet to be done to any significant degree. When viewed through the lens of Romantic texts as "radical pastoral" (Garrard), the multitude of allusions and direct references to Wordsworth, Clare, and other Romantic poets in *The Wind in the Willows*, as well as the often Romantic cast of *The Secret Garden*, look less like "childhood reminiscence visions of a rhetorical sublime"[86] and more like a method of encapsulating and camouflaging resistance to the environmental depredations of Judeo-Christianity, as well as "the restoration of a humaneness in the danger of killing divisions from home and work."[87]

The second ecocritical insight is John Parham's important proposal of a "Victorian ecology" that undermined societal norms more powerfully than its Romantic progenitors. In Parham's opinion, this was a world-view informed not only by the observed contrast between "pastoral" and urban-industrial ways of living, but also by "proximity to the origins of contemporary ecology"[88] and, as significantly, by the advent of Darwin's theory of evolution by the process of natural selection.

Jointly, these two ideas were blows to ideas of human superiority and separateness that give Grahame and Burnett's evocation of "pagan" belief systems much greater force than the pre-Darwinist Romantic poets' similar references to Pan, Triton, the nymphs, etc. Holly Blackford cites *The Secret Garden* as "unconsciously commemorating an influential myth of seasonal renewal that is over two thousand years old," namely the myth of Demeter and, more generally, "the Eleusinian Mysteries [which were] the precursor to Christian belief in the immortal soul ... and reconcile goddess or nature worship with Christianity."[89]

Where many critics perceive this presence of classical myth as a crutch in the transition from Christianity to secularism, we can interpret it just as validly as Victorians and Edwardians seizing eagerly on a belief system that accorded with their own natures and biophilic essence. As Joseph Beach observes, "Historically considered, the faith based on this concept of nature ... was Christianity freed from various superstitious, irrational and, so to speak, parochial elements, and yet retaining enough emotional force to satisfy man's craving for a moral and intelligible universe."[90]

This is all by way of saying that the presence of Pan in Grahame and Burnett's books is far from innocuous or incidental. The challenge to Western orthodoxy posed by an earth-oriented, animistic, cycle-observing belief system was at that point no longer the paper tiger that it had been, for the most part, in the early 1800s or even the later Romantic period: it was part of the imminent overthrow of the Christian belief systems that had nominally governed the West for some hundreds of years.

A Friend with horns and cloven hooves: Pan in *The Wind in the Willows*

The first edition of *The Wind in the Willows* had as its cover illustration the figure of Pan, goat-footed and bearded, horns curved back like a ram's and eyes slanted under wickedly arched brows. He is playing his pipes over the curled, sleeping forms of Rat and Mole. The extent to which it is hard to imagine such a cover on a children's chapter book today, especially in America, exposes the openly transgressive origins of modern children's literature in these "Golden Age" texts. It also supports Beverly Lyon Clark's contention that these radical components have been papered over (literally, in the case of *The Wind in the Willows*, with other less Satan-like images) to live a "secret, subterranean life."[91]

It would perhaps seem an overstatement to say that the book's Pan chapter casts a paganistic umbrella over the entire book, especially since his image now rarely or never features on the cover and has been rendered successively less diabolically in inside illustrations. The book remains disjointed and episodic, and the chapter "Piper at the Gates of Dawn" continues to be read and felt as more anomalous than characteristic. However, its origin as a unique, at times verbatim insertion into *The Wind in the Willows* from Grahame's philosophical-religious *Pagan Papers*, rather than the set of letters to his son from which most of the rest of the book was drawn, makes it representative of the worldview within which Grahame wrote the book, much as The Lord of the Rings' appendices, inserted from *The Silmarillion*, have little direct bearing on the book's action but are of supreme importance in establishing its enveloping non-Christian cosmology.

Even if Pan has been driven off the cover and into the text and illustrations, He still resides in the book, facilitating the syncretic "religion" that I propose exists in children's literature and which *The Wind in the Willows* has played such a large part in establishing and validating. David Rudd sees many subtle and more overt ways in which Pan "operates in Grahame's text like a pandemic. He can be detected at levels from the phonological to the thematic, and, between these, he is captured in both imagery and symbol; more indirectly, Pan's presence is also marked by the characters' own levels of awareness of him." [92] At the very least, this deity's explicit and awe-inspiring presence prevents reader and critic from plausibly putting a Judeo-Christian construction on the book—the chapter "Dulce Domum," in which Mole and Rat sing Christmas carols, notwithstanding.

"The Piper at the Gates of Dawn" takes Rat and Mole away from their boys'-adventure excursions with Toad and Badger and sees them searching for Little Portly, Otter's young son, who has been missing for days. Passages simultaneously portray Little Portly as human and as an animal susceptible to human mechanisms. He is, in Miles Kimball's view, "a Romantic child whose playfulness depends on transgression, but whose transgression leads both to physical dangers and to the risk of adult awareness of those dangers."[93]

Mole refers to Little Portly as a child, and Rat reports that the young otter hasn't learned to swim well yet, something that takes him well out of the nature of otters, who in the continual company of their mother learn to swim expertly at about two months. At the same time, the Otter family's main concern is for Little Portly's susceptibility to the weir on the river, as well as "traps and things—*you* know."[94] The blending of humans and other animals and the vulnerability of the young of the species are highlighted in this way, eliciting an identifying anxiety for the child reader over Little Portly's welfare.

The blending continues throughout the chapter, not least in the way Grahame's writing casts the mythological interrelatedness of the Greco-Roman human-animal figures as something approaching a scientifically validated expression of ecological interdependence. Pan's presence and the ecological science in the chapter show the incorporation of sublimity into a matrix of ecology in which nature is not seen by human eyes but nevertheless something in which humans are immersed and with which they are one. The landscape is described not only as a landscape, but as a system of nature, with niches and their inhabitants: "the hedges, the hollow trees, the runnels and their little culverts, the ditches and dry waterways"; the "reeds and rushes and osiers."[95]

The chapter's opening passages, while providing plentiful evidence for those seeking to assign the chapter a Romantic heritage, also contain a scientific and ecological specificity that reads for a stretch like a natural-history text. Grahame describes the willow-wren who opens the chapter and has "hidden himself in the dark selvedge of the river bank"[96] in a species-faithful fashion; the evocative and accurate descriptions of sounds made by "the busy little populations who were up and about"[97]; and the articulation of the break of dawn: "The horizon became clearer, field and tree came more into sight, and somehow with a different look; the mystery began to drop away from them. A bird piped suddenly, and was still; and a light breeze sprang up and set the reeds and bulrushes rustling."[98] With its matrix of sensory events, it is a description that anyone who has been out in a clear summer dawn will recognize.

The often concrete and literal tenor of Grahame's writing leading up to and away from the mystical-rhapsodic pages of the Pan encounter lines up with John Parham's observations about the differences between Romantic and Victorian ecological writing—two modes that are often conflated, especially in children's literature. Late Victorian ecological writing such as Grahame's, he points out, often contained "a conjunction of natural history and evolutionary theory, two paradigms widely acknowledged as the origin of ecological science."[99] These paradigms, however, were "troublesome" to early Victorian poets and of course generally unavailable to poets such as Wordsworth and Clare, who so evidently influenced Grahame's outlook.

Grahame's particular use of Pan and a seemingly Romantic mode of writing in the chapter reflects the enormous Darwinian paradigm shift that recontextualizes and radicalizes the Romanticism of many early giants of children's literature. The human understanding of the world that Grahame depicts both through humans and through anthropomorphized animals, as well as the understanding that he himself was writing from, was beginning to move away from the nature-segregating, nature-demoting Christian point of view even as society moved further into the technoindustrial dispensation based on that segregation and demotion.[100] Although critics have a strong case in claiming that Grahame alludes to and emulates the Romantic poets in this chapter and the book overall, the fact remains that the world-altering publication of *The Origin of Species* (1859) and *The Descent of Man* (1871), and the embattled ascendancy of evolutionary theory, stand between the publication of these poets' works and Grahame's book.

Intermixed with the chapter's "naturalism" is more than an intimation of immortality. Rat and Mole row onward toward Pan's seductive music in powerful terms that urge the child reader to a Damascene moment. As Horne and White observe, Grahame uses "diction that suits [the chapter's] religious topic":[101] floating foam-streaks" and the "solemn and soothing rumble" of the "shimmering arm-spread" of the weir lead to a revelatory island, "full of significance…hid[ing] whatever it might hold behind a veil, keeping it til the hour should come, and, with the hour, those who were called and chosen."[102] Rat and Mole, with whom the child reader has identified wholly and sympathetically, are being formally inducted into their holy order, allowed into the presence of their god. "'This is the place of my song-dream, the place the music played to me,' whispered the Rat, as if in a trance. 'Here, in this holy place, here if anywhere, surely we shall find Him!'"[103]

Thus, when Rat and Mole, having rowed up the river's backwater in pursuit of the elusive music they have heard, come upon Pan, he is felt as god to both the human and the animal child. In strikingly Christian vocabulary, Grahame terms this pagan figure "the Friend and Helper" whose "august Presence" paralyzes Mole with "an awe that smote and held him" and renders the usually authoritative Rat "cowed, stricken, and trembling violently."[104] Rat, "his eyes shining with unutterable love," says of his powerful response to Pan, "Afraid? Of *Him*? O, never, never! And yet—and yet—O Mole, I am afraid!"[105] Upon these words, "the two animals, crouching to the earth, bowed their heads and did worship."[106]

Yet, the object of this Rapture-like description appears as a benign Satan, with a "backward sweep of curved horns," "bearded mouth," "splendid curves of the shaggy limbs" and cloven hooves—the hallmark of deviltry in Christian mythology—upon which lies, "sleeping soundly in entire peace and contentment, the little, round, podgy, childish form of the baby otter"[107] On the one hand, one could hardly find a more archetypal depiction of a Christian revelation experience. On the other, one could hardly find a more overt depiction of devil-worship, or at the very least, idolatry and the flagrant breaking of the first and most important Commandment.

With this in mind, we would do well to pause and consider the purported origins of children's literature as religious Tractarianism and moral indoctrination. It is worth noting how remarkable it is that one of the unchallenged giants of the genre indirectly encourages sentiments that, if taken out of context, resemble devil-worship, and *in* context constitute an even more subversive idolatry: the nature-worship that early Christianity fought and sacrificed much to supplant. In my view, the extent to which this content has gone uncensored and even seemingly unnoticed speaks to two important facts. The first is the persistence and water-like pervasiveness of biophilia. I would argue that this is what directs adults toward fostering their young's connection with the natural world through literature if no other means are available. The second is the resulting inurement to nature-endorsing messages in children's literature even when they clearly present blasphemy to the Judeo-Christian paradigm.

Underscoring this open challenge to monotheism and Abrahamic religious doctrine, Grahame names Pan a "kindly demi-god" and bequeaths him the power "to bestow on those to whom he has revealed himself in their helping...the gift of forgetfulness,"[108] so as to protect them from growing miserable after "looking with mortal eye on things rightly kept hidden."[109] In making Pan a demi-god, rather than the full god that he was in the Greek belief system, Grahame creates another point of intersection: Pan, as a satyr already half-human and half-animal, is now also half-mortal and half-divine. He thus contains a simultaneity and a co-existentiality of the human, the animal, and the divine. This elevates the potentially secular ecological focus of the book to a religious plane, completes the link between the human world and the rest of the natural world, and relocates the site of reverence from the Christian Heaven to the biological matrix. The text suggests that we understand the sameness of human and animal sublimity, and that we see not God *in* nature, as natural theology would have it, but god *as* nature, the basic tenet of biotheism.

The song that Rat fleetingly recalls, after Pan has ushered in the oblivion that relieves those in his presence of paralyzing awe, depicts some humans as a scourge on the animal world, setting "traps and snares" that render "limbs...reddened and rent," creating "wounds [Pan] binds."[110] Such implications, though they do not openly accuse adult human society, leave little room for any interpretation a critic might make that Grahame was trying to argue for the sublimity of Man in a typically Romantic way. Instead, with Pan as the immortal presence, salving human-made wounds and saving small animals from limb-rending traps, the suggestion is that the reader reorient her or his own perceptions of what is reverence-worthy from human-centered and -arrogating religious constructs to the awesome realities of their natural surroundings.

Thus, the pagan images and spirituality portrayed in "The Piper at the Gates of Dawn" signify not merely a nostalgic hearkening back to a Classical, pre-Industrial agrarian idyll, but part of an assault on the fundamental tenets of Christianity. The Pan chapter redirects the child reader's sense of worship to the animalistic. In doing so, it inserts into the founding sensibilities of modern-day children's literature a suggestion that the Christian worldview, with its uniquely nature-defying concept of human immortality, was a temporary aberration in belief systems, falling between, on the one side, the human-as-animal view proposed pre-scientifically by Greco-Roman mythology, and, on the other, the scientific, post-Darwinian, secular "paganism" of the era that Grahame and Burnett were writing in.

The Magic-worker: Dickon as Pan in *The Secret Garden*

The interpretation of *The Secret Garden*'s Dickon as a Pan figure has been made as an obvious one, and has been casually accepted, within children's literature criticism. Seth Lerer, cross-referencing the "Pan" scenes in his annotated edition of *The Wind in the Willows* with *The Secret Garden*, writes, "Compare ... the scene in Burnett's *Secret Garden* when Mary Lennox comes upon Dickon: 'A boy was sitting under a tree, with his back against it, playing on a rough wooden pipe. He was a funny-looking boy about twelve' and all of the animals draw 'near to him to watch him and listen to the strange low little call his pipe seemed to make.'"[111]

Likewise, Fred Inglis repeatedly gives Dickon the appositive identity of "the Pan-boy,"[112] while *Secret Garden* annotator Gretchen Holbrook Gerzina assumes the identification throughout, despite otherwise assigning persistently Christian readings. Even before Dickon appears, Gerzina observes, "Burnett is drawing a picture of [him] as a boy who is most at home in nature ... He can also be seen as Pan, the god of woods, pastures, shepherds, and flocks."[113] And Burnett herself wrote of Dickon as "a Faun who charms wild creatures and tame ones."[114]

Although Dickon is described as a plain, ruddy-faced boy, with a turned-up nose and round eyes, Mary and Colin credit him with being a worker of "Magic," and Burnett reinforces this secondary characterization with considerable repetition and variations on the theme. Mary tells Colin that Dickon is "not like anyone else in the world. He can charm foxes and squirrels and birds just as the natives in India charm snakes."[115] Through him, Burnett and her characters begin to redefine the supernatural concept of "Magic" as the natural concept of unity with the living world. Dickon himself doesn't use the word Magic, as Mary explains to Colin, but simply feels as if he were a bird or an animal. And Colin, who begins as a sickly "anti-Dickon," convinced that he will die with any physical activity or exposure to fresh air, declares after entering into Dickon's domain, "I shall get well! I shall get well! … I shall get well! And I shall live forever and ever and ever!"[116] As Holly Blackford says, Dickon "integrates his mothering of animals and Mary into his pantheistic being."[117]

As well, fulfilling the ideas of metamorphosis sketched by Grahame's use of Pan in *The Wind in the Willows*, Dickon proposes a high degree of inter-identity between humans and other animals. He tells the nesting robin, within earshot of Mary, "Us is near bein' wild things ourselves. Us is nest-buildin' too."[118] Jennifer Marchant notes that such exchanges "emphasize the deep and long-standing connection between the two species, and hint that the interspecies harmony readers see in the garden is possible elsewhere."[119]

The transmission of Pan's life-force through Mary and from Mary is articulated when Colin says that Dickon is the only stranger he can bear to have look at him. He declares that Dickon is an animal charmer, and he himself is "a boy animal.'"[120] Though both Colin and Mary laugh at this description and ostensibly find the notion of a boy animal hiding in his hole amusing, it is clear that Burnett is working from the literality that these child characters, like all children, are the young of an animal species, and what is good for any other young animal—outdoor play, air and sunlight, healthy nourishment, deep sleep—is good for them.

Dickon, in his Pan-like way, imparts a sense of immortality to Colin. In one of the key phrases of the novel, Mary throws open the windows to Colin's stately tomb and urges him to absorb the immanence of nature: "That's fresh air … Lie on your back and draw in long breaths of it. That's what Dickon does when he's lying on the moor. He says he feels it in his veins and it makes him strong and he feels as if he could live forever."[121]

It is helpful here to enter into the counterfactual in order to appreciate fully the impact of this choice. Had Burnett wished to propound a Christian model of human-nature relationship, she could have written Dickon as a St. Francis of Assisi character, such as figures prominently, for instance, in Robert Lawson's *Rabbit Hill*. This would have encapsulated the nominally pagan, pantheistic, and biotheistic elements of the story within an overarching Christian rubric, neutralizing their heterodox implications as C. S. Lewis does in his Narnian chronicles by positioning the Christ-figure Aslan as lord over the Greco-Roman fauns, centaurs, naiads, and dryads. Similarly, she could have had Dickon repent of his overly animalistic, unschooled lifestyle, or at least show an interest in conventional behavior and learning equal to Mary and Colin's interest in the Rousseauean "education" of the outdoors. She could have had him request or receive this kind of schooling from his more formally educated and conventionally socialized new friends, as Heidi's friend Peter does in Johanna Spyri's classic. This would have put *The Secret Garden* in the realm of the natural theology of which *Heidi* is the flagship text and to which, rather confoundingly, some critics draw explicit comparisons.

By instead giving Dickon a pipe to play, and making this a conspicuous aspect of his ability to "work Magic" and "charm animals," Burnett instead clearly marks him as representing the *pre*-Christian immortal, the single deity in the Greco-Roman belief system who is most intimately associated with nature, wildness, and fertility. Further, Dickon remains the ideal for both childhood and adulthood, his deep knowledge of nature held up as the only education that really matters.

This choice helps shape the novel's non-Abrahamic conception of humans' position within and relationship to the natural world, and it creates a forceful resistance to any readings of the novel as propounding a Christian morality—or even operating within an Abrahamic paradigm. Choosing a pagan immortal both directs and defines Mary and Colin's identical character goals: not to acquire a set of Christian virtues and reform their willful ways, but to discover their animal natures and connect through them to the immanence of the growing, ever-renewing matrix of life. I would argue that the deliberately and conspicuously pagan frame created by the Dickon-as-Pan depiction lays the groundwork for quite a different morality. It is one resonant with the reincarnationist outlook in Oscar Wilde's long poem *Panthea*, his rebuke of and correction to Wordsworth's Ode:

> We are resolved into the supreme air,
> We are made one with what we touch and see
> ... ...
> ...[T]hose argent breasts of thine will turn
> To water-lilies; the brown fields men till
> Will be more fruitful for our love to-night,
> Nothing is lost in nature, all things live in Death's despite.[122]

Entering into immanence: Mary's story

A good deal of the syncretic, biotheistic religion in *The Secret Garden* involves the way it reforms the conception of immortality. Specifically, it shifts immortality from the linear Christian paradigm, in which life and death are binaries and the immortality of an afterlife is a third state physically and metaphysically removed from the other two, to a naturalistic conception of dormancy and renewal, with immortality derived from the reincarnation/recycling that absorbs us into nature's immanence.

This reformation begins in the first chapters, with the omnipresence of death making the concept of immortality both more concrete and more urgent. The book's opening chapter, the bleakly titled "There is No One

Left," starts the story off with the death of both of Mary Lennox's parents and countless Indians, who have been "dying like flies"[123] in a cholera epidemic. Mary does not enter the book as an orphan, as so many characters in children's literature do: she becomes one, graphically and brutally, before the reader's eyes. This scene of widespread death leaves Mary in what Cadwallader terms "a sort of post-apocalyptic isolation."[124] Entirely forgotten, and with her parents dead, Mary scrounges on leftover party food and sleeps alone in her nursery, wondering where everyone is. When she is eventually discovered by a British officer, she is told "There is nobody left to come"[125]—and it is "in this strange and sudden way," the narrator reinforces, "that Mary found out that she had neither father nor mother left; that they had died and been carried away in the night, and the few native servants who had not died also had left the house as quickly as they could get out of it."[126]

Mary herself is portrayed as flirting with death. She begins the novel as a listless little girl with a thin face, thin body, thin hair, and skin that is yellow because she has "always been ill one way or another."[127] This description has been read, and validly so, as part of a stereotypical conception of India and other hot climates as enervating and unhealthy. However, Mary's development as a fretful, ugly baby and toddler is also set in the context of a busy, office-centered, indoor human world in which Mary knows nothing except for the dark indoors and the faces of her Ayah and other servants.

Early intimations of nature as a ward against sickness come when Mary, left to her own devices amid the panic and sickness, goes outside and play-acts gardening, sticking broken-off hibiscus flowers into heaps of soil. Her futile emulations of life-force and revivification, with the conspicuously colored flowers acting as a talisman, foreshadow the later suggestion, and subsequently the open assertion, that connecting with nature physically and materially can keep us alive.

During her short stay with an impoverished English clergyman before returning to England, she similarly tries to create a garden, piling up heaps of earth, making paths, and earning herself the sobriquet "Mistress Mary, quite contrary." Mary's habitual garden-making, as practiced in her temporary home, is positioned as her lifeline to life-force, the one "wick" taproot in the otherwise sterile and enfeebling colonial culture. Here, with the explicit presence of a Christian adult, the two modes of mortality and immortality compete for the first but not the last time in the book.

Once at Misselthwaite Manor, on the Yorkshire moors, Mary is again confronted with death, this time in the form of her uncle Archibald Craven's dead wife Lilias—a conspicuous beauty, like Mary's mother. The crucial difference is that Lilias created and tended a garden while alive, her line to immortality, while Mary's beautiful, social-gadfly mother, connected only to ephemeral human events, is rarely mentioned again and dies a complete death, physically and literarily. Lilias Craven (unnamed until the final chapter), who loved her walled garden and spent much of her days there, lives on in it: her immortality is real. Mary connects with Lily Craven through the garden, so that the dead woman becomes something of a spiritual mother. Mary was born the year that Lily Craven was killed and the garden locked; in her act of secretly unlocking and then rejuvenating the garden, she acts as reincarnation, as well as daughter. Burnett thus offers the child reader a way to conceive of immortality as a literality, when the "Magic" of the garden begins to incorporate the idea of living "forever and ever."

The book begins its gradual exchange of death as a corporeal finality for a non-Abrahamic understanding of a cycle that includes periods of dormancy before rejuvenation once Mary first sets foot inside the walled, locked, "secret" garden. Mary wonders if the plants are all dead and wishes that it not be so. Accustomed to death's finality despite her youth, she has only a dim sense of the dormancy that mimics death but is a fundamental part of nature's immanence. "All that troubled her was her wish that she know whether all the roses were dead, or if perhaps some of them had lived and might put out leaves and buds as the weather got warmer. She did not want it to be a quite dead garden. If it were a quite alive garden, how wonderful it would be."[128]

This marks the beginning of Mary's (and later Colin's) lesson in the eternality of nature. It is a lesson that continues and is reinforced throughout the book, as one of the main Christian precepts incorporated into a non-Christian paradigm and transformed. In Holly Blackford's view, Mary at this point starts participating in "the story of hope and rejuvenation for mankind … a compromise between mother [Demeter] and consort [Hades] enables the seasonal cycling of Persephone between [earth and underworld]."[129] Although there are potentially Christian allusions in Mary's thoughts about her "secret kingdom,"[130] the configuration transmutes these tropes to a pre-Christian or non-Christian paradigm. The linear timeline of birth, death, and an absolute polarity of these two states has been re-bent into a cycle of dormancy and renewal.

Immortality thus retreats out of the realm of morality (Heaven) and into the realm of the literal and the active. No sooner has Mary begun wondering about the garden's state than she finds plants that she knows from the gardener, Ben Weatherstaff, to be alive—snowdrops, daffodils, crocuses. That evening, she anxiously asks of her servant, Martha, "Do bulbs live a long time? Would they live years and years if no one helped them?", and Martha assures her, "[M]ost of 'em will work away underground for a lifetime an' spread out an' have little 'uns."[131] Mary learns, too, about reciprocity without dependency, and about one aspect of the human species' ecological niche: the bulbs have lived without anyone weeding them, but they will thrive when she weeds them and gives them room to "breathe" amid the grass.

Such moves to subsume Christian allusions into a naturalist reframing are part of the syncretism of children's literature to which I refer earlier. These passages of *The Secret Garden* are, in my view, at the vanguard of transfiguring Christian doctrine into a *de facto* paganism that connects children from the late Victorian and Edwardian ages forward with a deeper, millennia-old biophilia, a cultural heritage that finds its way even through the severest impediments.

"It's the Magic": Colin's story

While a number of critics—including Elizabeth Lennox Keyser, Heather Murray, and Beverly Lyon Clark—have observed that "Mary recedes in importance as the book progresses and Colin gains center stage,"[132] an ecocritical reading might instead suggest a cycle of transmission. Dickon and Martha initiate Mary's reconnection with the outdoors, and she passes the spark of nature-worship on to Colin.

In this view, Mary's story is not the only important one, and it is only our assumption that the book will follow a conventional novelistic form, focusing on one main character, that would lead us to see Mary as jilted or dismissed. Through the lens of biophilic ecocriticism, Mary first is fulfilled as an individual by her discovery and stewardship of the garden, then continues to grow in her affinity for the outdoors and the earth, and finally is fulfilled socially and culturally when she initiates Colin into the wonders of nature. She initiates him, as well, into her new understanding of natural immanence, in contrast to the polarized abstraction of Christian immortality. The secret garden, in Jen Cadwallader's view, "allows Colin to master the Spiritualist concept of death as a continuation of life, epitomized in his exclamation, 'I shall live forever and ever and ever!'"[133]

Concern over Mary's "silencing" has sidelined an interesting consideration: Why *does* Frances Hodgson Burnett tell almost exactly the same story twice? More importantly, why, given that this narrative gemmation makes up the whole book, has *The Secret Garden* become not only the most successful and enduring of her own works but, in many critics' views, "the most significant children's book of the twentieth century"[134]? We would do well to ask this question in consideration of the novel's strong biophilic and biotheistic essence.

Colin's journey is Mary's journey all over again, bumped up a notch in intensity. As soon as Mary has grown fully "wick"—a key term in *The Secret Garden*, a Yorkshire word usually used in the book as an adjective but also a verb that succinctly describes what Dickon does with the life-energy of the moor and Mary with that of the garden—the cycle begins again with Colin. Before their physical and spiritual reformation, both Mary and Colin have been described as "always being ill," with Colin being bedridden and widely expected to die. Where Mary has been made listless and weak by lack of invigorating contact with the elements, Colin has been enervated, not to say terrified, by being led to expect that he won't live past childhood and, if he does, that he will be a hunchback. Where Mary experienced the life-sapping enclosure of a British colonial child kept on the compound grounds, Colin is a willful shut-in, curtailing his few outings because he hates people staring at him, and refusing to take one forthright doctor's advice to go outside because he hates fresh air.

With the Rousseauean prescription of playing outdoors in the strong moorland winds, Mary goes from weak and bored to healthy and engaged, while Colin goes from bedridden and wheelchair-bound to running, doing boxing exercises, and planning a career as a professional athlete. Early on in their cousinly acquaintanceship, Mary tentatively offers the vocabulary of "the Magic" to which she has been exposed by watching the fakirs in India; Colin embraces and formalizes it, enacting nature-worshipping rites to "call" the Magic to him. Mary eventually expresses a desire for Dickon's approval and affection; Colin immediately idolizes him and ascribes to him a special access to and adeptness with the "Magic" that will confer on him a form of eternal life.

Readers have presumably embraced this reiteration, as *The Secret Garden* is one of the best-loved and bestselling children's books yet written. It should therefore be noted that the main function of Colin's story is to amplify and reinforce each element of the novel's ecocentric message. Within an ecocritical analysis, one could say that where Mary's story implies, Colin's explicates. Where Mary's demonstrates, Colin's defines. Where Mary's outlines, Colin's underscores. And it is in Colin's story that the novel's biophilia is extended and formalized into biotheism, subsuming ideas of the afterlife into one of atomistic botanical reincarnation and embedding in Burnett's precedent-setting syncretism a conception of human-nature relationship. This influential act of permission-giving can be seen in, among others, Laura Ingalls Wilder's *By the Shores of Silver Lake*, many of Arriety's rhapsodic observations of nature in the Borrowers series, and the strongly urged biotheism of *Goodnight, Mr. Tom*.

Colin's first function is to highlight the novel's biophilic intimations. The rapidity and readiness with which he receives the life-force of nature's immanence bespeaks *The Secret Garden*'s underlying subscription to, and its persistent promotion of, the beliefs and observations that make up E. O. Wilson's "innate tendency to focus on life and lifelike processes"[135]: "the quickness and decisiveness with which we learn particular things about certain kinds of plants and animals,"[136] that "to explore and affiliate with life is a deep and complicated process in mental development,"[137] and that "[C]onsciousness races ahead to master certain kinds of information with enough efficiency to survive … [and] submits to a few biases easily while automatically avoiding others."[138]

Even hearing about the garden second-hand, Colin shifts from an existential limbo in which he is neither alive nor dead, and wants neither to live nor die, to an animated child beginning to wick up the energies of the natural world through hearsay. He becomes excited, his eyes shining and growing in awe as Mary tells him the story of the locked-away garden. By the end of her rapturous account, he wants to see the garden—even though, he notes, "I don't think I ever really wanted to see anything before."[139] This reception is so remarkable that Martha exclaims that Mary must have bewitched him—a reference, Mary assumes, to "Magic" of the kind she had heard about in India but has (she believes at this point) been unable to perform herself. Some days later, Colin makes the connection between the natural life of the garden and his own life even more explicitly: "If I could get into it [the garden], I think I should live to grow up!"[140]

Hearing Mary's vivid descriptions of the garden, Dickon, the animals and the moor, Colin increasingly revives, drinking in the images. With her talk of the outdoors and her refusal to indulge in Colin's morbid self-conception, Mary brings him to life. Dickon works through her, the power of Pan widening his scope: when Colin tells Mary that a medical expert has said he might live if he put his mind to it and had the right attitude, Mary immediately presents him with Dickon as the antidote to non-living and the right person to put him in a better mood: "He's always talking about live things … He's always looking up in the sky to watch birds flying—or looking down at the earth to see something growing. Let us talk about living. Let us talk and talk about Dickon."[141] And after having started the visit with thoughts of death, illness, and other subjects not considered by adults to be normal childlike ones, Mary and Colin finish it laughing "so that in the end they were making as much noise as if they had been two ordinary healthy natural ten-year-old creatures."[142]

Colin seizes on the idea of immortality via fresh air and meditation on nature. From morbidity, he goes to the other extreme: an impulsive belief in an immortality rooted in the literal eternality of natural immanence, animism, and reincarnation. Burnett writes of his first view of the "magical" secret garden in terms of both rhapsody and rapture, trying to capture visual simultaneity, Joyce-wise, in impressionistic, breathless run-on sentences:

> And over walls and earth and trees and swinging sprays and tendrils the fair green veil of tender little leaves had crept and in the grass under the trees and the gray urns in the alcoves and here and there everywhere

> were touches or splashes of gold and purple and white and the trees were showing pink and show above his head and there were fluttering of wings and faint sweet pipes and humming and scents and scents. And the sun fell warm upon his face like a hand with a lovely touch.[143]

"One with what we touch and see": ecological intermingling

Colin's exclamation after "coming alive" upon his first sight of the garden, "I shall get well! I shall get well! ... And I shall live forever and ever and ever!"[144] has been read as an entry into Christian immortality. However, little in the text supports this. Colin has not changed from a morally unworthy person into a pious one. What little improvement has taken place in his imperious, self-centered character is kept a secret among the children, so as to preserve the secret of the garden. If anything, Colin's increasing health makes him even more arrogant and grandiose, convinced that he will be a great scientist and a champion athlete. The natural and "pagan" imagery far outweighs any that could be construed as Christian, and, as I have argued, what Christian allusions do exist are transmuted into figures of bacchanalia, nature-worship, and biophilic attunement. Colin's exclamation, which forms the title of Chapter Twenty, seems instead a response to two quite unChristian things: firstly, Dickon's pagan presence and influence; and secondly, the literal immortality that has been conferred on Colin's mother, Lily, through the relative eternality of her garden, where the flowers *do* live "forever and ever."

The non-Christian meaning of Colin's feelings is explicated by the narrator at the start of the succeeding chapter. Those feelings stem from a connection with the eternal elements of natural creation—a "pagan" concept of immortality that echoes *Panthea*'s lines "We are resolved into the supreme air / we are made one with what we touch and see": a kind of spontaneous incarnation or biotheistic coinherence. The novel's narrator makes this identification clear as the definition of Colin's immortality:

> [O]nly now and then one is quite sure one is going to live forever and ever and ever. One knows it sometimes when one gets up at the tender solemn dawn-time and goes out and stands alone and throws one's head far back and looks up and up and watches the pale sky slowly changing and flushing and marvelous unknown things happening until the East almost makes one cry out and one's heart stands still at the strange unchanging majesty of the rising of the sun—which has been happening every morning for thousands and thousands and thousands of years. One knows it then for a moment or so.[145]

This is not, to use Eric S. Rabkin's phrase, "immortality [as] a self-defeating fantasy, a desperate defense against death."[146] It is, on the contrary, an acceptance of a non-Christian, non-individualistic transfiguration and an absorptive inter-identity that rejects strict boundaries and segregation between species and even between individuals. The depiction of immortality in *The Secret Garden* echoes ecologist Neil Evernden's argument that the concepts of individual identity and boundary conflict with the fundamental ecological tenet of interdependency: "To the Western mind, 'interrelated' implies a causal connectedness ... But what is actually involved is a genuine intermingling of part of the ecosystem. There are no discrete entities."[147]

Therefore, Colin, Dickon, and Mary's *Panthea*-like perception of essential or peripheral intermingling with immanent nature as part of their literal immortality is not mere mysticism, nor even Deep Ecology, but, as Evernden observes, the eventual conclusion and basis of scientific ecology itself. In fact, it presciently expresses the "subversive tenet of ecology" that "what we take for granted to be a single organism controlled by a single set of genes may in fact be more properly thought of as a very dependent kind of symbiosis, a mutualism."[148] Thus, "the desire of some ... to deal only with the fragment of reality they term 'human' is nonsense."[149] Evernden writes of the bacterial flora that inhabit our guts and upon which we are dependent for our survival as one reason that interrelatedness should be taken literally, "not simply as an indication of causal connectedness. Where do you draw the line between one creature and another? Where does one organism stop and the other begin?"[150]

It seems likely that this paradigm became rooted in children's literature largely through its exemplification in *The Secret Garden*—even while, conversely, *The Secret Garden*'s status as a flagship novel in children's literature supports the idea of the genre as a "carrier" of the biophilic impulse. To the extent that "the subversive nature of ecology as a science rests on its assumption of literal interrelatedness, not just interdependence"[151] and *The Secret Garden* promotes the key ecological precept that "there is no such thing as an individual, only an individual-in-context, individual as a component of place, defined by place,"[152] the book participates in disrupting the techno-industrial, urbanized, capitalist, and individualistic beliefs that form the nominal social matrices of Western life. Mary's/Burnett's exhortation to "lie on your back and draw in long breaths of [fresh air]" in order to "live forever and ever"[153] speaks to a mode of immortality that accords with, and indeed is an aspect of, ecology—a perception of Evernden's "genuine intermingling" and thereby a sharing in natural immanence.

The biotheistic conception of immanence in *The Secret Garden*, and the children's rapturous embrace of it, also refutes Bernard Williams' contention, as summarized by John Fischer, that in order to make immortality "truly appealing," there must be "a future in which an individual can recognize himself or herself—someone genuinely *identical* to the individual, not just qualitatively similar or with several identical properties."[154] This criterion of "identity condition" is a particularly narrow and largely Western-Abrahamic one. It precludes the ecological "intermingling" of *The Secret Garden*'s immortality and invalidates non-Western perspectives on the relationship between humans and their natural surroundings. The science of ecology has shown that these perspectives are scientifically more accurate than the supposedly empiricism-descended (perhaps actually Christianity-descended) scientific notions that gave rise to the technological and individualist views that nominally govern modern society.

The success of *The Secret Garden* also challenges Williams' attendant "attractiveness condition," which asserts that "the types of fusion envisaged in nonatomistic models (even ones that somehow preserve individual streams of consciousness) do not allow individuals to look forward to *their own* future existence. As such, these nonatomistic models are not very appealing models of immortality."[155] Evidently many millions of child readers, over many generations, have found the non-individualistic depiction of immortality/immanence in *The Secret Garden* "very appealing," and the adults who have bought books for children have felt the same.

In *The Secret Garden*, "immortality" comes from Magic—the magic of the garden, where Colin's mother's spirit seems to be active. Her presence, though, is written as natural rather than as supernatural. Her legacy is one of long-lived, spreading, "immortal" roses, bulbs, and trees. As the gardener Ben Weatherstaff tells the children, Colin's mother was "[n]ot one o' them as looked down on th' earth—not her. She just loved it."[156] This rejects the Heaven-focused abstraction of Christian paradise, with its promised afterlife. Although Lily Craven also liked skyborne things, she created her own paradise and her own afterlife through imbuing the earth of the garden with bulbs that could undergo the cycle of dormancy and awakening indefinitely, and perennials whose very name marks and demarcates the point on the yearly cycle, "dying back" to the ground in late fall and reemerging every spring.

"Magic," in turn, merges with science, again anticipating and confirming the ecological science of our current age, when empiricism has come full circle to "discovering" many of the "mystical" precepts of intermingling, reincarnation, and transmission it had originally rejected as baseless superstition. For Colin, "Magic" is the same as the "scientific experiment" he launches. Until recently in scientific history, his statement "The great scientific discoveries I am going to make … will be about Magic,"[157] might have seemed paradoxical. Indeed, an unsympathetic rationalist reader might view Colin, and Burnett with him, as naïve. However, time has shown that

Colin speaks wisely when he conflates the "Magic" of an assumed substantial intermingling—a botanical transubstantiation—with other elements of the local ecological web, and when he accepts Mary's and Dickon's assertion that the life-force of spring can infuse his blood and make him well.

Consolidating the syncretism

That Burnett makes numerous "religious" references is undeniable. Near the end of the book, these move into some openly Christian ones, such as when the children and Ben Weatherstaff, coached by Dickon, sing the Doxology. I would argue, however, that consistently, throughout the book, Christian ways of regarding humans' status and function within the larger environment, as well as Abrahamic beliefs in creation and mortality, have been composted, as it were. In their broken-down, particularized form, they provide elements of reverence, awe, and belonging, along with some taxonomies of divinity, to the nature-worshipping whole. The "Magic"-calling that Colin equates to the Doxology has been far more firmly established by that point in the book than any Christian worldview. In fact, with Mary's dubbing Colin a "young Rajah" and her references to fakirs, their rituals have from the start been removed from a Judeo-Christian context and aligned with the pantheistic, reincarnative beliefs of Hinduism. Mary feels enraptured by these rites, which Colin chants while the children and Ben Weatherstaff sit on the ground in a circle. Colin's incantation, as he stands in the garden looking like a priest, is given in full in the text, and he repeats it many times over—"not a thousand times but a goodly number."[158] After Dickon introduces Colin to the Doxology, Colin openly equates and conflates his incantations with religion, saying that the Doxology may express what he means when he wants to shout out his gratitude to "the Magic."

Burnett develops her novel's nature-worship with great elaborateness and sincerity in the closing chapters of *The Secret Garden*. Colin continues to fragment and mix boundaries and belief systems when he observes that everything is made out of "Magic" and lays out for the reader a detailed program for making the mystical cosmology work psychosomatically. He authoritatively tells Mary, Dickon, and Ben that the way to learn things like the Magic is to repeat them over and over and think about them "until they stay in your mind forever."[159] This repetition, he believes, will bring the Magic to our aid and make it part of us.

Susan Sowerby explicates and consolidates the novel's essentially biotheistic message when she tells the children about the underlying principle of these rituals, terming the combination of Magic, Hinduism, fakirism, Christianity, and pagan nature-worship "the Big Good Thing":

> [W]hat does th' name natter? I warrant they call it a different name i' France an' a different one i' Germany. Th' same thing as set th' seeds swellin' an' th' sun shinin' made thee a well lad an' it's th' Good Thing. It isn't like us poor fools as think it matters if us is called out of our names. Th' Big Good Thing doesn't stop to worrit, bless thee. It goes on makin' worlds by th' million—worlds like us. Never thee stop believin' in th' Big Good Thing an' knowin' th' world's full of it—an' call it what tha' likes.[160]

The idea that we contain "worlds" ourselves and that our world—the larger, external world, presumably—is only one of "millions" directly contradicts the Christian creation myth described in Genesis. In the process, it undoes the hierarchy at the root of monotheistic Abrahamic belief and the "chain of Being" concept that developed throughout the early and middle Christian eras.

The syncretic form of reverence fashioned explicitly here, and more implicitly in *The Wind in the Willows*, is so complete (and, perhaps not coincidentally, so similar to those of certain biospiritual "indigenous" ontologies) in regard to the relationship between humans and the natural world that it would not seem a stretch to view it as the autochthonous belief system of the British Isles showing through its thin Christian overlay. Books such as *The Secret Garden* and *The Wind in the Willows* suggest this overlay to be an anomalous and imposed worldview, a graft that failed to take culturally and psychologically, even if it prevailed economically. These novels' success in establishing tropes and eco-spiritual values that quickly came to dominate the genre furthermore offers the idea that children's literature from the mid-Victorian period onward has acted as a powerful, perhaps central, mechanism of rejection.

In other words, books like these, ingested by the young of the human species during the generations in which industrial capitalism, urbanization, and loss of contact with the land intensified, can be seen to have formed a Teflon layer for the new dispensation, even to the extent of thwarting the technological determinism that major philosophers have accepted as true. It is presumably both a cause and an effect of *The Secret Garden*'s influence and canonicity that this openly other-religious cant elicits little or no comment from the guardian class. Adults seem to turn a blind eye to the almost inevitably radical implications of texts for children stimulating a belief system, value system, and species self-perception that not only resists but actively contests those of Western cultural orthodoxy.

Transmitting biotheism adult-ward

The Secret Garden demonstrates the transmission of biophilia from children to adults in a mirroring of the way, I contend, children's literature does in modern Western societies. In other words, both the child characters in these two texts and the texts themselves serve as the receptacles, conservers, and eventually transmitters of "nature-endorsing" values and self-perceptions. The children in Burnett's novel teach/remind its adult characters to be aware and induct them into a biotheism-based syncretism.

Ben Weatherstaff comes to understand the synonymy of the Doxology and nature-worship through Colin's daily Magic-calling rituals, while Susan Sowerby is reported to have learned her deep "earth mother" wisdom from raising and observing her 12 nature-loving children. One of the things she learns is that children need fresh air and exercise—and also that children learn from their connection with nature the inevitability and importance of self-limitation. Susan herself learned at an early age that "th' whole orange doesn't belong to nobody ... What children learns from children ... is that there's no sense in grabbin' at the' whole orange."[161]

This distinctly ecological—not to say socialist—piece of wisdom comes up the chain from her semi-wild moor children to Mrs Sowerby. It is then transmitted across to less-attuned other adults in the novel when it is absorbed and sanctioned by the heir to the estate, the "young rajah," Colin. Non-Christianity resides and flourishes in, and ultimately is championed and insisted upon by, the heir to the initially Christian, allopathic Misselthwaite Manor, its fortune, and its position of authority in the region.

Colin's discovery/declaration of "immortality" is twinned with a similar awakening in the current authority, his father, hundreds of miles away in the Austrian Tyrol. After years apathetic and emotionally numbed by his beloved wife's death and son's invalid status, Archibald Craven finally begins to transcend his surname when he lets his mind and gaze rest on some brookside forget-me-nots. Taking in these symbols of natural immanence, he undergoes a spiritual transformation and murmurs, "I almost feel as if—I were alive!"[162] Across the miles, as the "Magic" of the garden steadily infuses his son with natural eternal life, Craven slowly comes back to life, becoming calm and gradually rising out of the black depression that has held him for years.

This links the botanical life-force with the human, and gives the child reader a sense of what Oscar Wilde in *Panthea* terms "mighty waves of single Being."[163] Archibald Craven returns to Misselthwaite, and from the abstraction born of self-severing from the land and its inhabitants due to his wife's death, he feels "he was on earth again"[164]—brought back to life by meditating his way back into his body and connecting with the ground, rather than pursuing the thought-driven life on which Western culture puts a premium.

We can view this as a stereotypical instance of "child as teacher," or, in Wordsworth's words, "the child being father of the man." However, we can also see it as the book's content mimicking its own sociocultural transmission: children's love for it (one born, in my view, of an affinity for the natural world it describes and, even more, its overt promotion and affirmation of that affinity) leads adults to obtain it for children and continue to publish and read it. This provides the adults with a pretext for treasuring a subversively ecocentric point of view and keeping that value set alive and vibrant within themselves. Beverly Lyon Clark has detailed how female adult readers in her informal survey cite *The Secret Garden* as "the secret love," aiding a historical rise in popularity that Clark sees as "even more subterranean" than that of *Little Women*.[165]

An ecocritical examination would pose rather different questions than those that Clark does regarding this matter. Where Clark quotes Anne Scott MacLeod as saying that "the Burnett book that passed from child to child, now as then, is not *Fauntleroy* but *The Secret Garden*,"[166] I would raise the point that for the most part, the novel's readership is younger than the age of purchasing independence, and therefore the children themselves are not passing it along. *Adults* are buying it, ostensibly for the children, and giving it to the children, thus keeping themselves in the "nature-endorsing" value loop.

This supports my argument that much children's literature acts as a subversively conservative element in modern Western society. In the case of *The Secret Garden*, it does so by clearly disseminating ecocentric values and impulses upward from the young of the species to the older, providing a way for adults to honor their earlier allegiance to nature and to affiliate with other animals in the living world. Clark's description of the novel's "grass-roots popularity"[167] is not only apt but perhaps more telling than she intended. These primeval inclinations toward animality, nature-worship, and a perception of life and death as an immortal cycle of dormancy and renewal are the evolutionary roots of our supposedly urban, techno-industrial, nature-deficient Judeo-Christian culture.

As well, where Clark inquires "why ... *The Secret Garden* [was] for so long such a well-kept secret,"[168] I would find it less important to ask why its popularity was "subterranean" than why it existed at all. Why did this book achieve and maintain a level of popularity, apparently entirely organic, that has endured and grown for 103 years, so that as of the mid-1990s, it continued to outsell even *Charlotte's Web* on publishers' backlists? Meanwhile, *The Wind in the Willows* "has been issued in more than 100 editions."[169] It has estimated *annual* sales of 80,000 copies and upward, more than 100 years after first publication. *The Secret Garden*, likewise, has been translated "into almost every living language ... [and] continues to sell three hundred to four hundred thousand copies a year."[170] Both have been the subject of centenary critical compilations: Kenneth Grahame's *The Wind in the Willows: A Children's Classic at 100*

(2009) and Frances Hodgson Burnett's *The Secret Garden: A Children's Classic at 100* (2011). Both have also been issued as hardcover, gift-edition annotated versions.

It may be just as telling that the two less nature-focused texts of the "Golden Age's" precedential quadrumvirate, *Alice in Wonderland* and *Peter Pan*, proved a barren line. While, as Gillian Avery enumerates, *Alice* had a number of imitators, including Alice Corkran's *Mrs Wishing-to-be*, Edward Knatchbull-Hugessen's *Ernest*, and Maggie Browne's *Wanted—a King*,[171] none of these endured, and it is rare today to hear of a children's book being "just like *Alice*" or "a worthy successor to *Peter Pan.*"

By contrast, canonical texts with a family likeness to *The Wind in the Willows* and *The Secret Garden* not only abound but positively predominate. *The Wind in the Willows* helped situate a two-directional theriomorphism-anthropomorphism at the heart of the canon, paving the way for successors such as Thornton Burgess's Mother West Wind's Children series, Robert Lawson's Newbery–winning *Rabbit Hill*, Robert C. O'Brien's *Mrs. Frisby and the Rats of Nimh* (also a Newbery Award winner), Sheila Burnford's *The Incredible Journey*, E.B. White's *Charlotte's Web*, Brian Jacques' Redwall series, and Katherine Applegate's Endling series, to name just a few. *The Secret Garden* helped to initiate and license the strain of children's literature whose narratives and themes center on independent children thriving through interactions with the natural world. This heritage can be seen in books ranging from Laura Ingalls Wilder's Little House series and Astrid Lindgren's *Pippi Longstocking* to Michelle Magorian's *Goodnight, Mr. Tom* and Suzanne Collins's dystopian-survivalist Hunger Games series.

If there were ever a datum to support the contention that children's literature sustains itself on biophilia, thriving on themes of nature and acting as Western society's chief source of cultural/imaginative connection with the population's evolutionary impulses, it is in the fact of *The Secret Garden* and *The Wind in the Willows* achieving the status of classics apparently through these forces alone. One could surmise that their stories create a receptacle for biophilia, providing joy and affirmation of the impulse toward nature for children. And then, through their popularity, they provide an excuse for adults not only to pass them along to children but to stay in touch themselves with this "contrary" part of their nature, supported in their internal rebellion by the "contrariness" of the characters who reject anthropocentric priorities and insist on expressing their animality and pagan-based "immortality."

I contend that these two giants of children's literature demonstrate two vital aspects of the genre vis-à-vis biophilia. First, in normalizing what I have termed the "syncretism" of children's literature, they opened the gates early and wide in the genre's history for open declarations of an ecologically transgressive view of the world. Secondly, their organic and increasing popularity suggests that children for the past 100-plus years, and the adults they became, long for and love these affirmations of identity with nature.

The Secret Garden and *The Wind in the Willows* have preserved this wisdom. Along with hundreds of equally nature-focused children's books, they have transmitted it safely through the hardening and straitening of both secular science and Abrahamic fundamentalism, into an era where the understanding of human interdependency, intermingling, and inter-identity has legitimized it.

Notes

1 Gillian Avery, *Nineteenth Century Children: Heroes and Heroines in English Children's Stories 1780–1900* (Hodder and Stoughton, 1965), 81.
2 Marah Gubar, *Artful Dodgers: Reconceiving the Golden Age of Children's Literature* (Oxford University Press, 2009), 3.
3 David Sandner, *The Fantastic Sublime: Romanticism and Transcendence in Nineteenth-Century Children's Fantasy Literature* (Greenwood, 1996), 4.
4 Humphrey Carpenter, *Secret Gardens: A Study of the Golden Age of Children's Literature* (Faber, 1985), x.
5 Carpenter, *Secret Gardens*, 120.
6 Sarah Gilead. Qtd in David Rudd, "*Deus ex Natura* or Nonstick Pan? Competing Discourses in Kenneth Grahame's The Wind in the Willows," in *Kenneth Grahame's The Wind in the Willows: A Child's Classic at 100*, eds. Jackie C. Horne and Donna White (Scarecrow Press, 2011), 4.
7 Seth Lerer, *Children's Literature: A Reader's History, from Aesop to Harry Potter* (Chicago University Press, 2008), 254.
8 Gubar, *Artful Dodgers*, 3.
9 Samuel F. Pickering, Jr., *John Locke and Children's Books in Eighteenth-Century England.* (University of Tennessee Press, 1981), xi.
10 John Rowe Townsend, *Written for Children: An Outline of English-language Children's Literature* (Penguin, 1983), 40.
11 Jean-Jacques Rousseau, *Emile: Or, On Education* (The Floating Press, 2009), 196.
12 Lawrence Buell, *The Environmental Imagination: Thoreau, Nature Writing, and the Formation of American Culture* (Belknap/Harvard University Press, 1995), 85.
13 Jane Darcy, "The Representation of Nature in *The Secret Garden* and *The Wind in the Willows*," *The Lion and the Unicorn* 19, no. 2 (1995): 211.
14 Darcy, "Representation of Nature," 214.
15 Miles Kimball, "Aestheticism, Pan, and Edwardian Children's Literature," *Critic* 65, no. 1 (2002): 50.
16 Jackie C. Horne and Joe Sutliff Sanders, eds., *Frances Hodgson Burnett's* The Secret Garden*: A Children's Classic at 100* (Scarecrow Press, 2011), xxix.
17 Jackie C. Horne and Donna White, eds., *Kenneth Grahame's The Wind in the Willows: A Child's Classic at 100* (Scarecrow Press, 2010), xxxvi.
18 Horne and White, *Kenneth Grahame's The Wind in the Willows*, xxviii.
19 Horne and White, *Kenneth Grahame's The Wind in the Willows*, xxi.

20 Horne and White, *Kenneth Grahame's The Wind in the Willows*, xxiii.
21 Darcy, "The Representation of Nature," 214.
22 Frances Hodgson Burnett, *The Secret Garden* (Harper, [1911] 1990), 203.
23 Kenneth Grahame, *The Wind in the Willows* (Yearling, [1908] 1974), 9–10.
24 Grahame, *Wind in the Willows*, 126.
25 Michael Martin, *Albert Schweitzer's Reverence for Life: Ethical Idealism and Self-Realization* (Ashgate, 2007), 32.
26 Martin, *Albert Schweitzer*, 33.
27 Carpenter, *Secret Gardens*, 13.
28 Jen Cadwallader, "The Three Veils: Death, Mourning, and the Afterlife," in *Frances Hodgson Burnett's The Secret Garden: A Children's Classic at 100* (Scarecrow Press, 2011), 120.
29 Gretchen Holbrook Gerzina, *Frances Hodgson Burnett: The Unexpected Life of the Author of* The Secret Garden (Rutgers University Press, 2006), 6.
30 Gerzina, *Frances Hodgson Burnett*, 6.
31 Albert J. Kuhn, "English Deism and the Development of Romantic Mythological Syncretism," *PMLA* 71, no. 5 (1956): 1094.
32 E. O. Wilson, *Biophilia* (Harvard University Press, 1984), 48.
33 Alun Morgan, "Places of Transformation in *The Secret Garden*," in *Frances Hodgson Burnett's The Secret Garden: A Children's Classic at 100* (Scarecrow Press, 2011), 82.
34 Fred Inglis, *The Promise of Happiness: Value and meaning in children's fiction* (Cambridge University Press, 1981), 112.
35 Holly Blackford, "Maiden, mother, mysteries: the myth of Persephone in *The Secret Garden*," in *Frances Hodgson Burnett's The Secret Garden: A Children's Classic at 100* (Scarecrow Press, 2011), 3.
36 Inglis, *The Promise of Happiness*, 112.
37 Cadwallader, "The Three Veils," 121.
38 Townsend, *Written for Children*, 89.
39 Blackford, "Maiden, Mother, Mysteries," 15.
40 France Hodgson Burnett, *In the Garden* (Medici Society of America, 1925). Qtd in Horne and Sanders, *Frances Hodgson Burnett's*, 2.
41 Carpenter, *Secret Gardens*, 2.
42 Deborah Dysart-Gale, "Techne, Technology, and Disenchantment in *The Wind in the Willows*," in *Kenneth Grahame's The Wind in the Willows: A Child's Classic at 100*, eds. Jackie C. Horne and Donna White (Scarecrow Press, 2010), 25.
43 Dysart-Gale, "Techne, Technology, and Disenchantment," 24.
44 Dysart-Gale, "Techne, Technology, and Disenchantment," 24.
45 Carpenter, *Secret Gardens*, 21.
46 Morgan, "Places of Transformation," 82–83.
47 Carpenter, *Secret Gardens*, i.
48 Garrard, "Radical Pastoral?" 182–83.
49 Peter Gould, qtd in John Parham, "The Poverty of Ecocritical Theory: E. P. Thompson and the British Perspective," *New Formations: A Journal of Culture, Theory and Politics* 64 (2008): 157.
50 Darcy, "Representation of Nature," 213.
51 Darcy, "Representation of Nature," 213.
52 Darcy, "Representation of Nature," 213.
53 Gould, qtd in Parham, "The Poverty of Ecocritical Theory," 158.
54 Wilson, *Biophilia*, 1.
55 Dysart-Gale, "Techne, Technology, and Disenchantment," 23.

56 Grahame, *Wind in the Willows*, 1.
57 Kathryn V. Graham, "Of School and River: *The Wind in the Willows* and its Immediate Audience," *CLAQ* 23, no. 4 (1998): 181.
58 Horne and Sanders, *Frances Hodgson Burnett's The Secret Garden*, xxiii.
59 Gretchen Holbrook Gerzina, *The Annotated Secret Garden* (Norton, 2007), 23.
60 Gerzina, *Annotated Secret Garden*, 119.
61 Burnett, *Secret Garden*, 198.
62 Burnett, *Secret Garden*, 213.
63 Townsend, *Written for Children*, 89.
64 Morgan, "Places of Transformation," 82–83.
65 Morgan, "Places of Transformation," 82.
66 Rousseau, *Emile*, 11.
67 Rousseau, *Emile*, 33.
68 Morgan, "Places of Transformation," 81.
69 Morgan, "Places of Transformation," 84.
70 Burnett, *Secret Garden*, 127.
71 Burnett, *Secret Garden*, 169.
72 White, Jr., "Historical Roots," 7.
73 Rudd, "Deus Ex Natura," 9.
74 Patricia Merivale, *Pan the Goat-God: His Myth in Modern Times* (Harvard University Press, 1969), viii.
75 Rudd, "Deus Ex Natura," 9.
76 Kimball, "Aestheticism, Pan," 4.
77 William Greenslade, "'Pan' and the Open Road: Critical Paganism in R. L. Stevenson, K. Grahame, E. Thomas and E. M. Forster," in *Outside Modernism: In Pursuit of the English Novel, 1900–30* (Macmillan and St. Martin's Press, 2000), 146.
78 Lerer, *Children's Literature*, 156.
79 Beverly Lyon Clark, *Kiddie Lit: The Cultural Construction of Children's Literature in America* (Johns Hopkins University Press, 2003).
80 Madelon S. Gohlke, "Re-Reading *The Secret Garden*," *College English* 41, no. 8 (1980): 894–902. Qtd in Cadwallader, "The Three Veils," 122.
81 Kimball, "Aestheticism, Pan," 4.
82 Kimball, "Aestheticism, Pan," 6.
83 Greenslade, "'Pan' and the Open Road," 146.
84 Sandner, *The Fantastic Sublime*, 3.
85 Garrard, "Radical Pastoral?", 182.
86 Lerer, *Children's Literature*, 22.
87 Inglis, "The Promise of Happiness," 114.
88 Parham, *Environmental Tradition*, 156.
89 Blackford, "Mother, Maiden, Mysteries," 18.
90 Joseph Warren Beach, *The Concept of Nature in Nineteenth-Century English Poetry* (Pageant, 1956), 5.
91 Clark, *Kiddie Lit*, 30.
92 Rudd, "Deus Ex Natura," 9.
93 Kimball, "Aestheticism, Pan," 7.
94 Grahame, *Wind in the Willows*, 117.
95 Grahame, *Wind in the Willows*, 120.
96 Grahame, *Wind in the Willows*, 116.
97 Grahame, *Wind in the Willows*, 119.
98 Grahame, *Wind in the Willows*, 121.

99 Parham, *Environmental Tradition*, 157.
100 White, Jr., "Historical Roots," 9.
101 Horne and White, *Kenneth Grahame's* The Wind in the Willows, xxii.
102 Grahame, *Wind in the Willows*, 123.
103 Grahame, *Wind in the Willows*, 123.
104 Grahame, *Wind in the Willows*, 125.
105 Grahame, *Wind in the Willows*, 126.
106 Grahame, *Wind in the Willows*, 126.
107 Grahame, *Wind in the Willows*, 126.
108 Grahame, *Wind in the Willows*, 127.
109 Grahame, *Wind in the Willows*, 126.
110 Grahame, *Wind in the Willows*, 130.
111 Lerer, *Children's Literature*, 148.
112 Inglis, "The Promise of Happiness," 112.
113 Gerzina, *Annotated Secret Garden,* 39.
114 Gerzina, *Annotated Secret Garden*, 132.
115 Burnett, *Secret Garden*, 173.
116 Burnett, *Secret Garden*, 255.
117 Blackford, "Mother, Maiden, Mysteries," 16.
118 Burnett, *Secret Garden*, 197.
119 Jennifer Marchant, "'A Real Person—only Nicer': The Robin as a Companion Species," in *Frances Hodgson Burnett's The Secret Garden: A Children's Classic at 100*, eds. Jackie C. Horne and Donna White (Scarecrow Press, 2010), 63.
120 Burnett, *Secret Garden*, 184.
121 Burnett, *Secret Garden*, 235.
122 Oscar Wilde, *Panthea*, in *Poems*, n.p, accessed January 7, 2012, Bartleby.com.
123 Burnett, *Secret Garden*, 5.
124 Cadwallader, "Three Veils," 123.
125 Burnett, *Secret Garden*, 8.
126 Burnett, *Secret Garden*, 8.
127 Burnett, *Secret Garden*,1.
128 Burnett, *Secret Garden*, 95.
129 Blackford, "Mother, Maiden, Mysteries," 4.
130 Burnett, *Secret Garden*, 99.
131 Burnett, *Secret Garden*, 99.
132 Clark, *Kiddie Lit*, 31.
133 Cadwallader, "Three Veils," 121.
134 Sanders and Horne, *Frances Hodgson Burnett's The Secret Garden*, xxiii–xiv.
135 Wilson, *Biophilia*, 1.
136 Wilson, *Biophilia*, 1.
137 Wilson, *Biophilia*, 1.
138 Wilson, *Biophilia*, 101.
139 Burnett, *Secret Garden*, 158.
140 Burnett, *Secret Garden*, 216.
141 Burnett, *Secret Garden*, 177.
142 Burnett, *Secret Garden*, 177.
143 Burnett, *Secret Garden*, 255.
144 Burnett, *Secret Garden*, 255.
145 Burnett, *Secret Garden*, 265.
146 Eric Rabkin, "Immortality: The Self-Defeating Fantasy," in *Immortal Engines: Life Extension and Immortality in Science Fiction and Fantasy*, eds. George

Slusser, Gary Westfahl, and Eric S. Rabkin (University of Georgia Press, 1996), xvi.
147 Neil Evernden, "Beyond Ecology: Self, Place, and the Pathetic Fallacy," in *The Ecocriticism Reader: Landmarks in Literary Ecology*, eds. Cheryll Glotfelty and Harold Fromm (University of Georgia Press, 1996), 95.
148 Evernden, "Beyond Ecology," 94.
149 Evernden, "Beyond Ecology," 95.
150 Evernden, "Beyond Ecology," 94–95.
151 Evernden, "Beyond Ecology," 102.
152 Evernden, "Beyond Ecology," 103
153 Burnett, *Secret Garden*, 255.
154 Martin Fischer and Curl, "Philosophical Models," 3.
155 Fischer and Curl, "Philosophical Models," 4–5.
156 Burnett, *Secret Garden*, 283.
157 Burnett, *Secret Garden*, 285.
158 Burnett, *Secret Garden*, 291.
159 Burnett, *Secret Garden*, 288.
160 Burnett, *Secret Garden*, 235–36.
161 Burnett, *Secret Garden*, 233.
162 Burnett, *Secret Garden*, 342.
163 Wilde, *Panthea*, n.p.
164 Burnett, *Secret Garden*, 351.
165 Clark, *Kiddie Lit*, 30.
166 Clark, *Kiddie Lit*, 30.
167 Clark, *Kiddie Lit*, 30.
168 Clark, *Kiddie Lit*, 30.
169 Everett, "The 20 Greatest Children's Books Ever."
170 Clark, *Kiddie Lit*, 30.
171 Avery, *Nineteenth Century Children*, 130.

Bibliography

Avery, Gillian. *Nineteenth Century Children: Heroes and Heroines in English Children's Stories 1780–1900*. Hodder and Stoughton, 1965.

Beach, Joseph Warren. *The Concept of Nature in Nineteenth-Century English Poetry*. Pageant, 1956.

Blackford, Holly. "Maiden, Mother, Mysteries: The Myth of Persephone in *The Secret Garden*." In *Frances Hodgson Burnett's The Secret Garden: A Children's Classic at 100*, edited by Jackie C. Horne and Donna White. Scarecrow Press, 2010.

Buell, Lawrence. *The Environmental Imagination: Thoreau, Nature Writing, and the Formation of American Culture*. Belknap/Harvard University Press, 1995.

Burnett, Frances Hodgson. *In the Garden*. Medici Society of America, 1925.

Burnett, Frances Hodgson. *The Secret Garden*. 1911. Penguin, 1960.

Cadwallader, Jen. "The Three Veils: Death, Mourning, and the Afterlife." In *Frances Hodgson Burnett's The Secret Garden: A Children's Classic at 100*, edited by Jackie C. Horne and Donna White. Scarecrow Press, 2010.

Carpenter, Humphrey. *Secret Gardens: A Study of the Golden Age of Children's Literature*. Faber, 1985.

Clark, Beverly Lyon. *Kiddie Lit: The Cultural Construction of Children's Literature in America*. Johns Hopkins University Press, 2003.

Darcy, Jane. "The Representation of Nature in *The Secret Garden* and *The Wind in the Willows*." *The Lion and the Unicorn*, 19, no. 2 (December 1995): 211–22.
Dysart-Gale, Deborah. "Techne, Technology, and Disenchantment in *The Wind in the Willows*." In *Kenneth Grahame's The Wind in the Willows: A Child's Classic at 100*, edited by Jackie C. Horne and Donna White. Scarecrow Press, 2010.
Evernden, Neil. "Beyond Ecology: Self, Place, and the Pathetic Fallacy." In *The Ecocriticism Reader*, edited by Cheryll Glotfelty and Harold Fromm. University of Georgia Press, 1996.
Everett, Lucinda. "The 20 Greatest Children's Books Ever." *The Telegraph* (U.K.), April 3, 2010. Accessed February 26, 2014. Telegraph.co.uk.
Fischer, John Martin and Ruth Curl. "Philosophical Models of Immortality in Science Fiction." In *Immortal Engines: Life Extension and Immortality in Science Fiction and Fantasy*, edited by George Slusser, Gary Westfahl, and Eric S. Rabkin. University of Georgia Press, 1996.
Garrard, Greg. "Radical Pastoral?" In *The Green Studies Reader: From Romanticism to Ecocriticism*, edited by Laurence Coupe. Routledge, 2000.
Gerzina, Gretchen Holbrook. *The Annotated Secret Garden*. Norton, 2007.
Gerzina, Gretchen Holbrook. *Frances Hodgson Burnett: The Unexpected Life of the Author of The Secret Garden*. Rutgers University Press, 2006.
Gohlke, Madelon S. "Re-Reading *The Secret Garden*." *College English* 41, no. 8 (1980): 894–902.
Graham, Kathryn V. "Of School and River: *The Wind in the Willows* and its Immediate Audience." *CLAQ* 23, no. 4 (Winter 1998): 181–86.
Grahame, Kenneth. *The Wind in the Willows*. 1908. Yearling, 1974.
Greenslade, William. "'Pan' and the Open Road: Critical Paganism in R. L. Stevenson, K. Grahame, E. Thomas and E. M. Forster." In *Outside Modernism: In Pursuit of the English Novel, 1900–30*, edited by Lynne Hapgood and Nancy L. Paxton. Macmillan and St. Martin's Press, 2000.
Gubar, Marah. *Artful Dodgers: Reconceiving the Golden Age of Children's Literature*. Oxford University Press, 2009.
Horne, Jackie C., and Joe Sutliff Sanders, eds. *Frances Hodgson Burnett's The Secret Garden: A Children's Classic at 100*. Scarecrow Press, 2011.
Horne, Jackie C., and Donna White, eds. *Kenneth Grahame's* The Wind in the Willows*: A Child's Classic at 100*. Scarecrow Press, 2010.
Inglis, Fred. *The Promise of Happiness: Value and meaning in children's fiction*. Cambridge University Press, 1981.
Kimball, Miles. "Aestheticism, Pan, and Edwardian Children's Literature." *Critic* 65, no. 1 (2002): 50–62.
Kuhn, Albert J. "English Deism and the Development of Romantic Mythological Syncretism." *PMLA* 71, no. 5 (December 1956): 1094–116. https://doi.org10.2307/460527.
Lerer, Seth. *Children's Literature: A Reader's History, from Aesop to Harry Potter*. Chicago University Press, 2008.
Lerer, Seth. *The Annotated Wind in the Willows*. Belknap/Harvard University Press, 2009.
Marchant, Jennifer. "A Real Person—only Nicer: The Robin as a Companion Species." In *Frances Hodgson Burnett's The Secret Garden: A Children's Classic at 100*, edited by Jackie C. Horne and Donna White. Scarecrow Press, 2010.
Martin, Michael. *Albert Schweitzer's Reverence for Life: Ethical Idealism and Self-Realization*. Ashgate, 2007.

Merivale, Patricia. *Pan the Goat-God: His Myth in Modern Times*. Harvard University Press, 1969.
Morgan, Alun. "Places of Transformation in *The Secret Garden*." In *Frances Hodgson Burnett's the Secret Garden: A Children's Classic at 100*, edited by Jackie C. Horne and Joe Sutliff Sanders. Scarecrow Press, 2011.
Parham, John. "Was There a Victorian Ecology?" In *The Environmental Tradition in English Literature*, edited by John Parham. Ashgate, 2002.
Parham, John. "The Poverty of Ecocritical Theory: E. P. Thompson and the British Perspective." *New Formations: A Journal of Culture, Theory and Politics* 64 (2008): 157.
Pickering, Jr., Samuel F. *John Locke and Children's Books in Eighteenth-Century England*. University of Tennessee Press, 1981.
Rose, Jacqueline. *The Case of Peter Pan, or the Impossibility of Children's Fiction*. Macmillan, 1984.
Rousseau, Jean-Jacques. *Emile: Or, On Education*. 1762. The Floating Press, 2009.
Rudd, David. "*Deus ex Natura* or Nonstick Pan? Competing Discourses in Kenneth Grahame's The Wind in the Willows." In *Kenneth Grahame's The Wind in the Willows: A Child's Classic at 100*, edited by Jackie C. Horne and Donna White. Scarecrow Press, 2010.
Sandner, David. *The Fantastic Sublime: Romanticism and Transcendence in Nineteenth-Century Children's Fantasy Literature*. Greenwood Press, 1996.
Townsend, John Rowe. *Written for Children: An Outline of English-language Children's Literature*. 2nd rev. ed. Penguin, 1983.
White, Jr., Lynn. "The Historical Roots of Our Ecologic Crisis." In *The Ecocriticism Reader*, edited by Cheryll Glotfelty and Harold Fromm. University of Georgia Press, 1996.
Wilde, Oscar. *Panthea*. 1881. *Poems*. Accessed January 7, 2012. Bartleby.com.
Wilson, E. O. *Biophilia*. Harvard University Press, 1984.
Wilson, E. O. "Biophilia and the Conservation Ethic." In *The Biophilia Hypothesis*, edited by Stephen R. Kellert and Edward O. Wilson. Island Press, 1993.

2 Explicating anomie

Drowned Ammet and *The Search for Delicious*

I have proposed that the pervasive presence of nature-oriented themes in Western children's literature makes it a key vehicle for smuggling subversive biophilic values down the generations. If we accept this proposition, then we might view Natalie Babbitt's *The Search for Delicious* (1969) and Diana Wynne Jones's *Drowned Ammet* (1977) as transparent Trojan Horses, revealing the normally hidden "enemy" within and suggesting the saboteur's role of children's literature.

Admittedly, it is common enough for children's books to feature child protagonists with a deep affinity for other living beings, as *The Search for Delicious* does with the dreamy orphan Vaungaylen and *Drowned Ammet* does with the neglected peasant boy Alhammitt (Mitt). The ubiquity of the nature-loving protagonist in children's literature is a cornerstone of my argument. But these two novels do something that most other children's books do not: they explicitly position childhood biophilia against a nature-spurning adult society, making this conflict of values and priorities their main theme and the choice between the two paradigms the crux of Gaylen and Mitt's character arcs. By showing the incompatibility between the biophilic urges fostered by much Western children's literature and the worldview of industrialized, urbanized, Christianized adult society, *The Search for Delicious* and *Drowned Ammet* draw attention to the potentially radical function of nature-endorsing children's literature. In this sense, the two novels could be said to play my thesis as an open hand.

Drowned Ammet is the second and arguably most overtly Deep Ecological of Jones's mythopoeic Dalemark Quartet. The series re-imagines an alternative late-medieval Britain (Dalemark) as an animistic land whose people are beginning to forget or repudiate the Undying deities of earth, water, and harvest as they engage in a long civil war and move slowly toward industrialism. In parts of the heavily oppressed southern half of Dalemark, the Undying have historically been conceived in a manner that strongly resembles pre-Christian Celtic pantheism, with the "Ammet" of the book's title being the god of sea and harvest, and his mate, Libby Beer,

DOI: 10.4324/9781003784210-3

being the goddess of fertility and creation. The indigenous people who dwell in the archipelago called the Holy Islands venerate the two deities, referring to them by the names "The Earth-Shaker" and "She Who Raised the Islands." However, for most South Dalemarkers, "Poor Old Ammet" and Libby Beer now exist merely as straw and waxen effigies that every year are thrown into the ocean as part of a festival ceremony whose purpose and origins have been forgotten.

Growing up in this setting of vestigial nature-worship largely eradicated by class politics, the young peasant boy Mitt early on feels a powerful pull toward a scent-filled, nature-immersed "perfect place" that he believes to be his true home. However, in a re-enactment of the country's history, Mitt finds his intuitive sense of inter-identity with nature obscured by human-centered concerns after his parents move from the countryside to a port town, where Mitt gets involved with a group of freedom fighters trying to overthrow the aristocratic class. Fleeing for his life after a failed attempt to assassinate the repressive Earl Hadd during the festival of "drowning" Old Ammet and Libby Beer, Mitt eventually arrives in the Holy Islands, and recognizes them as his long-sought "perfect place." There he encounters an embodiment of Old Ammet (originally Alhammitt, the source of Mitt's own name), renews his allegiance to the land, and recovers/discovers his true identity as part of the undying ecological divine.

The Search for Delicious, in its creative origins intertwined with Babbitt's better-known *Tuck Everlasting* (1975), similarly runs a political plotline parallel to an ecological theme. On the face of it, Babbitt's book, with its nominally lighthearted tale of mermaids, dwarfs, and magical whistles, would seem to be the less profoundly ecological of the pairing in this chapter. Indeed, under Greg Garrard's "positional" headings, *The Search for Delicious* looks like a close fit with the relatively "shallow" category of "environmentalism," whose members are "concerned about natural resource scarcity or pollution but would look to governments or non-governmental organisations ...to provide solutions."[1]

On further examination, however, it seems clear that this "shallowness" stems more from the nascent state of Western environmental thinking at the time of the book's publication than from Babbitt's own views. Babbitt clearly does not subscribe to what Garrard names as the "shallow" environmentalist's slate of beliefs in "liberal democracy, human rights, Christianity and notions of historical or scientific progress as valuable."[2] Both the novel's content, which "demands recognition of intrinsic value in nature,"[3] and its overt, sustained misanthropy instead mark *The Search for Delicious* as a work of "deep ecology"—according to Garrard the most influential of "the four radical forms of environmentalism."[4]

The novel juxtaposes and eventually interweaves the long-term ecological history of a small, mountain-ringed watershed/kingdom, whose human residents have gradually lost their belief in natural guardians/spirit inhabitants, with a trivial battle over which food should complete the metaphor "Delicious is ____" in the king's new dictionary.

The protagonist, the Prime Minister's nature-loving foster-son Gaylen, is sent on a countrywide survey to learn the will of the people on this matter, but gets caught up in a brewing civil war of economic self-interest among the land's food producers, as they become convinced that their livelihoods will suffer if the food representing "delicious" does not include their own product. As the political bickering intensifies, Gaylen discovers a greater threat, and a greater quest, in the king's brother-in-law's plot to block the source of the kingdom's river system and thereby control all the people of the kingdom. In the end he comes to understand that our most fundamental self-interest is a harmonious, sustainable relationship with the land and water that support us.

Jones and Babbitt use factional politics in a similar and specific way: to represent and encapsulate the Western "adult" paradigm that prioritizes humanity's short-term concerns at the expense of the people's spiritual-emotional access to and respectful treatment of the natural world. This employment of politics, which I will expound upon at greater length below, supports and illuminates several central aspects of my argument. To begin with, it clearly delineates childhood/biocentric and adult/anthropocentric value sets as separate and mutually incompatible entities, and it makes these two paradigms the chief bases of differentiation between the human characters. By using politics symbolically to position adult values as hostile to biophilic priorities, the two novels also underline the radicalizing effect of any social element that fosters these priorities, including children's literature.

As importantly, having an enclosed social arena to represent the adult world allows both authors to move their child characters back and forth across the paradigmatic border, making the transition both internal and reversible. In the process of moving from their original nature-embracing worldviews to a nature-spurning adult view and back again, Mitt and Gaylen suffer deep crises of self-alienation, anxiety, and confusion. Both boys finish the novels in a state of psychic schism which neither character is shown as resolving or fully recovering from. This is precisely the kind of alienation and internal anomie that I argue exists—in some measure through the active agency of children's literature—within the Western adult population at large.

Moreover, Jones and Babbitt show this psychological sundering as extending from Mitt and Gaylen being forced to choose between two of their most powerful obligate adaptive complexes. On the one hand, there is the biophilic impulse—in E. O. Wilson's words, "the urge to affiliate with other forms of life… [which] unfolds in the predictable fantasies and responses of individuals from childhood onward [and] cascades into repetitive patterns of culture across most or all societies."[5] On the other, there is the ultrasocial imperative, or humankind's unique species urge to practice extensive extra-kin cooperation, with the concomitants of a deeply rooted need to conform and a horror of expulsion, that, as sociologists speculate, may make humans new in the evolutionary history of cooperation.

Read from a sociobiological point of view, Mitt and Gaylen's characters are subtle but telling literary renderings of the conclusions reached by researchers such as Roger S. Ulrich, Gene Myers, Andreas Meyer-Lindenberg, and E. O. Wilson regarding the implications and effects of thwarting the biophilic instinct in order to maintain the social one. Namely, they believe that having to choose between these previously conjoined evolutionary developments comes at a high cost in unhappiness, addiction, psychiatric disorders, heightened existential stress, and maladaptive behavior.

Deep ecology and animism

Drowned Ammet and *The Search for Delicious* begin their service as exemplars of subversive biophilia by showing the openness with which—and the extent to which—children's texts are permitted to promote a worldview that is at once Deep Ecological (that is, an ontological outlook holding that "the well-being and flourishing of human and non-human life on Earth have value in themselves...[and that] these values are independent of the usefulness of the non-human world for human purposes"[6]), monistic (that is, viewing humans and other living beings as basically the same descriptively and evaluatively), and spiritually mystical. If extricated and explicated, such an ontology would be shown to challenge most of modern Western society's key cultural tenets, from Christianity's human-exceptionalist doctrines and the workings of techno-industrial capitalism to the rigid pronouncements of "scientism." Both *Drowned Ammet* and *The Search for Delicious* link power and wisdom with reverence for nature, folly and evil with disregard or contempt for it. Magic and immortality in *Drowned Ammet* and *The Search for Delicious* reside solely in nature and those beings inextricably linked to it; the novels' deities are, like D. H. Lawrence's, "of the earth rather than of the sky."[7]

Jones and Babbitt start off by describing their settings as ecosystems overlain by palimpsests of human habitation, in a parallel of the ecological conflict underlying the human civil wars in both books. The countries in both books are inhabited and guarded by animistic or representative beings. In the detailed ecological descriptions with which Jones opens *Drowned Ammet*, the narrator mentions that the New Flate, the fertile farmland on which Mitt spends his early years, has been converted from marshland by the grandfather of the current tyrant, Earl Hadd. This pairing of social and environmental domination is more than incidental in a book whose title character, once a god signifying the people's gratitude for farmland and sea, has become merely a straw object of vague superstition.

Meanwhile, on the occasion when Mitt, as a toddler, pursues the scent of his "perfect place" out of the New Flate and onto the Old Flate, he finds the reverse of this ecological "reclaiming": the ground here is described as wide, treeless, and covered with marsh plants. In places, there are signs of former dikes and farms, but these have been erased by nature doing its own reclaiming. Such layering of differing time scales, each providing its own perspective and pointing to both the persistence and the vulnerability of nature, is a fundamental element of the Dalemark Quartet. It is summarized perhaps most explicitly in the final book of the series, *The Crown of Dalemark*, where, a year after the events in *Drowned Ammet*, the modern-day girl Maewen travels 200 years back in time and arrives at the site of what will be a major city in her time. She observes,

> Where they had first stopped could have been the space which Kernsburgh Central Station was going to fill, except that the waystone was all wrong. Those low mounds were where she had last seen shops and office blocks, and the slightly higher hummocks ahead … were where the Tannoreth Palace would be some day soon. The green crease she was following, full of hoof prints and horse droppings, was probably King Street. And instead of cars and lorries, there was a much quieter confusion of riders in two different liveries. Maewen could so little believe this was really Kernsburgh that she had to look up towards the distant hills to make sure. There she saw the blue jagged shapes she saw from Dad's apartment, the North Dales Peaks. But the oddest part was the way there had obviously been a city here once, under all these lumps. She felt as if time had stood upside down and she really was in the far future, looking at the remains of the Kernsburgh she had known.[8]

The prologue of *The Search for Delicious*, meanwhile, describes at length the origins of Gaylen's kingdom as a watershed. The valley that the human characters of the book will eventually know as the kingdom began, the narrator informs us, as a dusty, arid place": "There were winds and dwarfs there, but no mermaids because there were no lakes, and … no woldwellers either because forests couldn't grow in so dry a place."[9] After a long period of time, the dwarfs lose a mountaintop spring, creating a lake that eventually overflows and turns the valley into a watershed with a network of streams that water the land and turn it green. A number of fantastical creatures come and occupy their species niches in a remarkably Darwinian way. The lake fills with water creatures, including "a lovely little mermaid"[10]; as the ecosystem turns from arid to verdant, forests arise and their guardians, the woldwellers, take up residence to watch over them. In the novel's world, the traditionally fantastic obey ecological laws, blending the supernatural with the scientific.

These backstories—explicated in *The Search for Delicious*, merely touched upon in *Drowned Ammet*, which has the prehistorical *Spellcoats* and the quartet's two other books to provide a more comprehensive long view of Dalemark's existence—allow the reader to see the terrain of the kingdom not in what E. O. Wilson calls the "*organismic time*" natural to a large-mammal species such as *Homo sapiens*,[11] but in "*ecological time*," where "ecosystems [rather than individuals] ... have become the creatures of our vision"[12] (emphasis in the original).

It is this scale, in both books, that casts immediate human political concerns as trivial and gives transience to even the most seemingly permanent edifices and historical events. In *The Search for Delicious*, people are, as a species, an *arriviste* group, not only unable to "stay in their own place" and keep that place beautiful, but apparently lacking a place to keep. By the time of the main action in book, the people have forgotten, if they ever understood it, their country's essential nature as a watershed. They take as coincidence the homeomorphic overlap of its geopolitical and ecological boundaries, and call it a kingdom because it is ruled by a king. By the same token, they have forgotten their dependence on the land and water upon which they have built their transitory kingdoms.

In *Drowned Ammet*, with its "reclaimed" and reverted marshlands, ancient half-forgotten rituals of nature-worship, and hinted-at histories and prophecies, it seems that humans did once have a place within the natural matrix, but they now have moved away from it and become alien to it. In both novels, people's separation from nature, their pretense of superiority to it, and their literal reliance upon it are all simultaneously conveyed through the authors' technique of ecological superimposition.

Paired paradigms: biocentric childhoods, anthropocentric adults

With their Deep Ecological bases established in the opening pages, *Drowned Ammet* and *The Search for Delicious* go about aligning—and, in the case of the former, even equating—their protagonists with nature. Mitt is characterized, in the opening pages, almost wholly in terms of his communion with nature. He grows up literally immersed in the landscape of the New Flate, running along the dike paths and teaching himself to swim by falling into a dike when his parents aren't looking (an early intimation of their criminal and perverted parental negligence, which develops into turning their son into a suicide bomber for their own political ends). The sound of the breeze over the Flate is intertwined in his early memories with his parents' laughter. Nevertheless, he has a stronger sense of a different place as home: a place where "it's all quiet, with water" (229). When Mitt is two or three and the breeze blowing off the sea dies down for the first time in his experience, Mitt senses, or rather, smells, a "perfect place":

> The smell was cow dung and peat and trampled grass, mixed with smoke from the chimney ... Beyond that was the smell of fresh things growing–cow parsley, buttercups, a hint of may, and strongest of all, the heavenlike scent of willows budding. While, at the back of it, there and not there, so that Mitt almost missed it, was the faint boisterous bite of the distant sea… . [I]t was a land. It was not far off, just beyond somewhere, and it was Mitt's very own.[13]

Young Mitt doggedly sets off to find the place while he can still sense it, but ends up hopelessly lost on the wide Flate. To the young army officer who rescues him, he insists that the place he's looking for is "his," and although he lives at Dike End, the other place is "where I come from, and I'm going to my home."[14]

The scene where Mitt is pulled toward his "perfect place" has several points of significance. First, it underscores the ecological divine at the core of Mitt's identity—for, as we learn in the final book of the quartet, Mitt is one of Dalemark's Undying, an ecological immortal whose main charge is to defend the lands and waters against human ensnarement. It is through this intimation of nature-based immortality that Jones makes the reader feel Mitt's subsequent subornation by human-centered politics as apostasy and self-betrayal. Second, it ends with Mitt trekking for miles over farmland and marshland to find the "perfect place," despite his being barely beyond the toddling stage. Children, as Jones portrays them in this book, not only can feel an intense identification with nature but have the will to act on it, if not always yet the means. The rescuing officer, in fact, is amazed at the sight of a toddler standing alone in the middle of the Flate. Third, it foreshadows the figurative and literal imposition of parents, politics, and adult priorities between Mitt and his intense biophilic urges, which amount to a sense of inter-identity (well-founded, as it turns out) with the elements of nature.

Mitt remembers his "perfect place" every time the wind dies down on the Flate, "but he did not set out to look for it again. It was plain to him that soldiers only brought you back if you went."[15] Some months later, when a vindictive tax collector is in the process of driving Mitt's parents off their land, Mitt's yearning for this "perfect place" intensifies. He wants it more than ever and longs to try again to find it, but by now he is old enough to know that his duty lies with helping his frivolous, narcissistic mother. Again and again, the adult world intervenes in Mitt's desire to reach the place of full biointegration that he senses and seeks.

Gaylen, too, is established as a sympathetic protagonist mostly through his affinity for nature. Within the us-and-them, nature-versus-human scenario laid out in the opening pages, Gaylen serenely takes for granted that he is an "us," and that no human concern could possibly be of anything other than secondary importance to him. Babbitt conveys this less by overt description than by making Gaylen a naive echo of the misanthropic narrator. He dismisses as "silly" any signs of a split between humans and the rest of the living world, including the growing foment over which food is the most "delicious." When his foster-father, the Prime Minister, foretells that this trivial issue will lead to civil war, social upheaval, and division within the kingdom's populace, and that ultimately people will be distracted from the truly important things—"[S]pring will escape notice! Little flowers will be pushed up, only to be trodden down, and birds will sing unheeded!"[16]—Gaylen thinks complacently, "[N]onsense! How could such a thing be?"[17] Upon learning that he is to be the king's emissary for the poll of "delicious," he looks out the tower window of the palace at the spring-greening land beneath stretching away to and prides himself that he "knows what the world is like."[18]

More importantly, Gaylen is portrayed as a "believer": an example of what the narrator approvingly names as "the occasional child" with the imagination to recognize as real the woldwellers, mermaids, and other stewards of the land—"to admit to the possibility," she tells us, "of something even more amazing in the world than the commonplace marvels which it spreads so carelessly before us every day."[19]

At 12 years old, Gaylen has been made aware that he should have outgrown this reverence and fascination with the living world. He pretends not to listen to the first town's mayor at nights, when the mayor tells his young daughter, Medley, "fairy" tales of wise fish, white deer, and haunted castles, along with stories of dwarfs, mermaids, and woldwellers. He tries to hide from the mayor how eagerly he is listening, ashamed that he still half-believes the stories, and that he wants to believe them wholly. (The child reader, of course, has learned from the book's prologue that these beings do exist, making this one of many instances of dramatic irony that casts the adults as foolish and irreverent.)

The mayor later chastises his daughter for saying that she had actually run into the woldweller the previous summer; clinging to this belief indicates a worrisome adherence to a belief system that clashes with the adult world's supposedly rationalist view. He warns her that he will stop telling her fairytales if she is going to start believing them, and later that evening he issues Gaylen a veiled threat that he, too, will be ostracized from the right-thinking adult domain if he allows himself to believe in these creatures: "'Children get ideas into their heads sometimes' … 'Of course, you're only a child yourself, really.'

'Oh, no,' Gaylen protested. 'I'm twelve years old.'"[20]

The mayor then goes on to extol the greater rationality of the modern-day people who have ceased to believe in the stewards of nature, telling Gaylen pompously of the time when silly people believed in animistic nature-guardians such as woldwellers. He scorns them as having "very little insight in their outlook,"[21] and goes on to claim smugly that modern citizens of the kingdom are "much too advanced" for "superstitions" of that sort.[22]

Since we, the readers, know from the Prologue that these creatures actually *are* real, we know that belief is the correct position, disbelief the wrong. It follows, then, that to become a grown-up is to become incorrect and force oneself into an attitude toward the natural world that does not accord with available knowledge, intuitive understanding, or the wisdom of oral tradition. It is, in fact, to wilfully blind oneself to humans' obvious dependence upon and integration with other living beings. From our position of dramatic irony, we can't help but urge Gaylen to keep believing. Unlike the injunction in *Peter Pan* to keep believing in fairies, though, this belief represents a fuller, more complex, and (paradoxically, given the folk-tale figures Babbitt uses) more scientifically accurate understanding of the ecosystem.

This is the first but not the last place where Babbitt drives home the idea that childhood perceptions, as well as being the more reverential, are also the more accurate and rational. Adulthood, conversely, is a descent into a silliness born of inattentiveness, misplaced incredulity, and an inability to reason forward from cause to consequence—attributes usually assigned to children.

Even as they establish as their novels' primary tension the question of whether the child characters will retain their ecocentric worldview or lapse into a perspective that considers only human society's vested interests, Jones and Babbitt position that human (adult) society as hostile to the ecocentric child's values. Although each book has at least one traditionally drawn individual villain, both texts clearly identify the true antagonist as being a short-sighted, self-involved Western human culture at large, and the true villainy as being humans' denial of their dependency on the earth.

In *Drowned Ammet*, South Dalemark's adult society has begun to scorn the ritual and forget the deities who represent the humans' dependence on land, sea, and weather. The long-standing harvest-time Sea Festival in the major port town of Holand, in which the straw-plaited figure of Poor Old Ammet and the fruit figure of his wife, Libby Beer, are thrown into the sea to drown is considered a joke:

> The procession that went with [the dummies] was both noisy and peculiar. When they reached the harbor, they said traditional words and then

> threw both dummies into the sea. Nobody knew why this was done. To most people in Holand the ceremony was just an excuse to have a holiday, eat sweets, and get drunk. On the other hand, everyone would have thought it horribly unlucky not to have held the Sea Festival.[23]

The sense that failing to hold it would be "terribly unlucky" is a holdover, as Jones would have us infer, of this tradition's former status as a cult of veneration for the animistic deities. The sense of "bad luck" might also stem from some atavistic memory on the people's part that, as we learn elsewhere in the quartet, the Undying must be rendered in image in order to stay "bound into godhead" (*The Crown of Dalemark* 275, *The Spellcoats* 215)—with a possible implication being that as human society forgets or denies them, they are free to forget and neglect human society.

Jones portrays the oppressed peasantry of South Dalemark hardly more sympathetically than she does the earls and their henchmen. The commoners, as much as the aristocracy, have ceased to recognize any of the Undying. Near the middle of *Drowned Ammet*, Mitt's revenant father, Al, tells Mitt and his two fugitive companions that in another part of South Dalemark, "they had a saying… that Holanders kept gods they didn't own to. And that's true. I bet you didn't know they were gods at one time."[24] Despite having spent considerable time in the Holy Islands, whose residents *do* refer to the beings as gods, Al maintains, just as he did when he was derisively giving Mitt his "dirt-common" name, that keeping effigies of Old Ammet and Libby Beer is just a peculiar superstition. Both the political oppressors and the politically oppressed fail to recognize the figures as the harvest god and fertility goddess that they clearly are and the "drowning" as a ritual of death and rebirth meant to remind people of their ultimate indebtedness to nature.

The people of *The Search for Delicious*, as well, have forgotten their fundamental dependency on nature. Any reverence they may once have had has been lost in their interminable squabbling. Like the scornful but still fearful Holanders in *Drowned Ammet* who observe the rituals of their gods without admitting their religion, the adults of the kingdom have stopped believing in the woldweller—"but at the same time," that creature tells Gaylen, "they're afraid I might be true, so they won't let their children come [into the forest]."[25] The woldweller considers human questions—even those from ancient days, when adults still did believe in him—as "nonsense," being about love, war, and how to gain power. (It is worth pointing out that the woldweller is as unwise in his disdain for humans as they are in theirs for him: Babbitt takes a page out of J. R. R. Tolkien's book and has the woldweller fail to understand the forest's dependence on humans, as Hemlock is devising plans to dam the lake whose river system allowed the forest to grow up in the first place.)

In fact, in its depiction of adult society, *The Search for Delicious* falls squarely into Scott Slovic's rhetorical category of "jeremiad,"[26] an "overtly political" mode of nature writing "the primary goal of which is to persuade its audience to adopt a new perspective by pointing out the problems with readers' current way of thinking."[27] The first paragraph of the book informs us,

> There was a time once when the earth was still very young, a time some call the oldest days. This was long before there were any people about to dig parts of it up and cut parts of it off. People came along much later, building their towns and castles (which nearly always fell down after a while) and plaguing each other with quarrels and supper parties. The creatures who lived on the earth in that early time stayed each in his own place and kept it beautiful.[28]

True to the book's lineage of both Cotton Mather and James Fenimore Cooper (for *The Search for Delicious* is nothing if not a primer for the themes explicated in *The Last of the Mohicans* and *The Pioneers*), Babbitt reverts to rhapsody when describing nature, using image-filled "elegant prose"[29] to create, in Barry Lopez's words, "an environment in which thinking and reaction and wonder and awe and speculation can take place."[30]

For instance, when Gaylen is daydreaming his way into the forest—or rather, he is absent-mindedly letting his wise horse take him there—the normally quick-moving narration gives way to rhapsodic nature writing of the first order: "[T]he rain had begun to fall. It plopped and plipped on the leaves over his head, and here and there occasional drops that found their way down through the branches fell like small stars to hang trembling in his hair … There was a lovely green glow in the forest, a glow pierced everywhere by tree trunks like fingers thrust into an aquarium full of tinted water."[31]

This shift of rhetorical register, which Babbitt performs consistently throughout the book, has the effect of turning the book into what fantasy theorist Farah Mendlesohn calls a "portal-quest" fantasy—the mode of fantasy writing in which "a character leaves her familiar surroundings and passes through a portal into an unknown place," almost always on a clearly stated quest. In this kind of fantasy, Mendlesohn posits, "the fantastic is on the other side and *does not leak*"[32] (emphasis in the original).

The "unknown place," in the case of *The Search for Delicious*, is intimated to be a combination of remaining old-growth forest and its original spiritual system, a different place with different values and, as it turns out, a different and parallel quest to the symbolically "silly" and trivial one of polling the kingdom for the most delicious food. The horse takes Gaylen through the portal, which is the forest. As he passes into it, surrounded by trees and the sound of rain on the leaves high above him, Gaylen feels "as if he were going deeper and deeper into a world that existed tranquil and quite separate from the one he had left behind."[33] This separate world makes sense of the belief system and mystical awe which in the other world has been disdained as a mark of childishness. By the time Gaylen is deep in the forest, the context makes him feel that the woldweller must indeed be real.

The suggestion here is significant: that the natural world is now an "otherworld." The deep harmony that children can access is on "the other side" of a breach, a sealed one that "does not leak" back into what adults consider the real world. It is across this divide that Gaylen goes and returns, both bodily and psychologically, each world carrying its own separate and conflicting value set. The use of jeremiad and portal in Babbitt's book enforces the notion that humans sit outside of the ecosystem not through notions of superiority or even in value-neutral ways, but in the negative way often invoked by deep ecologists, as implicitly having forfeited their position in it through their inclination to "dig parts of [the earth] up and cut parts of it off" rather than to "stay each in his own place and [keep] it beautiful." They are a species apart, "plaguing each other with quarrels and supper parties"[34] and "enjoying" a great many troubles, "all of which they created quite by themselves."[35] Within the framework of ecological time and the immanence of creation, humanity exists within short (and short-sighted) cycles of trivial bickering, coming and destroying and going, irritating the woldwellers with foolish questions about violence and power, and building towns and castles that inevitably crumble.

Politics as anthropocentric symbol

Having established the childhood/biocentric paradigm as positive and the adult/anthropocentric one as negative, *Drowned Ammet* and *The Search for Delicious* begin moving Mitt and Gaylen from one to the other by introducing them into the typically "adult" realm of factional politics. I want to point out here that the presence of factional politics is in itself highly unusual. In the genre of children's literature, and even until recently young-adult literature, politics per se—by which I mean politics closely defined, as dealing with class struggle, factions, policy-making, and modes of governance—has been until recently nearly taboo. Notwithstanding the considerable amount of subversive subtext that authors and illustrators of children's picture books have at times managed to sneak into their books, as Julia Mickenberg has documented in her fascinating book *Learning from the Left* (2006), the fact remains that one could read nearly every children's book currently and formerly in publication without coming across a mention of the Republican Party, Fox News, the U.K. Labour Party's perennial fight with the Tories, and so on. We can see this in the fact that although "political viewpoint" is a named category of objection in the American Library Association's "most-challenged books" lists, as of 2020, it was named in only 6.5 percent of challenges or bans,[36] and few of these were books written for children.

Most of the time this prohibition against politics is tacit, but occasionally it surfaces—as when, for instance, prominent children's literature scholar Perry Nodelman evinces equal horror at the idea of "children reading, say, pornography or *Doonesbury* cartoons."[37] This casual yoking suggests that the interdict on politics in children's literature extends from the concept of the child reader as someone who must be protected from any intrusion of "real life," and children's literature as existing "exactly in terms of adult concerns about how children need adult protection from violence and monstrosity."[38]

Whether this is the case or no, it would seem that the formless naif whose psyche will be scarred for life from reading *Doonesbury* is not the child reader Diana Wynne Jones and Natalie Babbitt had in mind when they wrote *Drowned Ammet* and *The Search for Delicious*. Mitt goes from being a nature-loving child to, for all intents and purposes, a suicide bomber. Gaylen is transformed from a naive, dreamy child to the lightning rod for the increasingly violent ire of the farmers, millers, herders, and other food producers who perceive the poll about the most "delicious" food as a potential threat to their livelihoods. Both boys receive graphic, firsthand lessons about matters of economic interest, theories of class disparity, the problems of factionalism, and the workings of demagoguery. More than this, their participation in political life puts their safety, and even their survival, in real jeopardy. Mitt stands to be tortured and hanged if Earl Hadd discovers his assassination plot. Gaylen endures first verbal abuse, then stone-throwing, and finally attempts to shoot him as the spat over "delicious" mounts to a full-blown armed conflict.

This "differently presupposed" child reader, as we may call it, presents a couple of significant implications for the relationship between children's literature and ecological literacy, which I feel is important to touch on here before moving on to the more immediate textual functions of politics in *The Search for Delicious* and *Drowned Ammet*.

The first is that it suggests that texts have an extensive and permanent impact on the child reader's psychology. This contrasts with traditional assumptions, especially those from psychoanalytical theory, which assume in children a "prodigiously porous" mind[39] through which textual material sluices with little lasting effect on the well-numbered stages of identity-ideation and life stages. In the Freudian, Lacanian, and even Piagetian conceptions, a child might ingest with relative immunity and impunity an endless number of ecologically themed texts that in any other context would be recognized as deeply influential, perhaps even indoctrinating.

The reader implied by Jones and Babbitt's texts, however, is a psychologically tough child who can handle "adult" realities, accept the fact of an insecure existence, make independent judgments, act on conviction, and discern between value systems. The idea of *this* reader taking in, for example, the deep misanthropy of *Bambi*, the powerful yearning for a Native American–modeled relationship with the land in *By the Shores of Silver Lake*, the indictment of deforestation in *Tuck Everlasting*, or the animistic spiritualism of *The Secret Garden* puts a different slant on ecocentrism in children's literature. It opens up the possibility that such books might affect a child reader's existential outlook with something approaching the same depth and permanence that *Walden* or *The Pioneers* might an adult reader's.

The second important implication lies in the way the use of politics enforces the point that Mitt and Gaylen's turn away from nature stems from external forces, not an innate "maturing" process. By depicting the characters' rejection of their initial "nature-endorsing" orientation as requiring active indoctrination, rather than as happening "naturally" as part of their development and growth, Babbitt and Jones support the biophilia hypothesis, presenting human allegiance to the matrix of nature as the normal, default state for both children and adults. This effectively reverses the respective value placed on ecocentrism and anthropocentrism within humanist-based critical paradigms. Where such approaches hypothesize the child-nature affinity to indicate the "pre-rational" state of the child, who is still ignorant enough to believe that "whatever has life has life very much like our own,"[40] Babbitt and Jones portray those who deny or suppress that affinity as, in a manner of speaking, "post-rational."

Textually, the more immediately radical effects of the politics in *Drowned Ammet* and *The Search for Delicious* stem, as I suggest at the beginning of the chapter, from the way the author uses class politics to stand in for the adult worldview as a whole. It is a worldview in which immediate self-interest and short-term thinking are particularly amplified and whose relationship to the non-human natural world ranges from self-absorbed indifference to uneasy disdain to exploitative contempt. The presence of this contrasting paradigm provides Jones and Babbitt a unique degree of manipulability in their novels. Typically, to the extent that the clash of nature-endorsing and nature-scorning worldviews is represented in children's literature at all, it is represented in one of two ways.

The first is the clash depicted as a coming-of-age tale, in which the child protagonist proves her or his "maturity" by demonstrating that a utilitarian view of animals and nature has permanently eclipsed a sense of affinity and identification with them. Examples include classics such as Robert Newton Peck's *A Day No Pigs Would Die* and Marjorie Rawlings Kinnan's *The Yearling*. The second typical mode of presenting the nature-vs.-anti-nature clash is by locating the anti-nature worldview within an adult antagonist, as in E. B. White's *Charlotte's Web*, Dick King-Smith's *The Sheep-Pig* (known in the U.S. as *Babe, the Gallant Pig*), Elyne Mitchell's *The Brumby*, Roald Dahl's *The Fantastic Mr. Fox*, and countless other animal tales with heartless adult characters. This way of handling the conflict allows the child protagonists themselves to remain morally and psychologically intact.

By instead containing the adult worldview within a well-delineated sphere of the novels' ontology, Jones and Babbitt can insert Mitt and Gaylen into "adulthood," show the scarring psychological effects of them abandoning their earlier allegiances and affinities, and extract them again. In the process, the authors can—and do—throw into relief the resultant confusion, internal conflict, and alienation from both ways of relating to the natural and human worlds. Read from this angle, the two novels support and illuminate a key tenet of sociobiology: that forcing people to choose between their need for social belonging and an access to the natural world that fulfills the human brain's need for the complexity, diversity, and environmental changeability within which we evolved—"not just the minimal contact needed to exist," as E. O. Wilson says, "but a luxuriance and excess spilling into virtually everything we do"[41]—has serious consequences in compromised mental health and existential impoverishment.

Gaylen enters politics through the seemingly innocuous means of going forth on horseback to poll the king's subjects on what is the most "delicious" food. In the ethical and characterological framework of the novel, this equates to an explicit turning away from the deep truths and meaningful living of nature-immersed life and toward a willful adoption of human trivial-mindedness, in the form of "squabbles." The two modes are portrayed as incompatible: as soon as people get caught up in political self-interest, Gaylen's foster-father the Prime Minister predicts, this ridiculous spat will overtake all other aspects of adult life in the kingdom, pushing out the meaningful elements of living and making people heedless of the dawning spring, with its flowers and birdsong.

In the condensed time-frame of the novel, which takes place over only about a week, Gaylen starts his poll about the most "delicious" food entranced with the natural world, but by the time he has left the first town's bellicose population and daydreamed his way into the part of the forest where the woldweller lives, already the cognitive dissonance has become a serious matter. Torn between awe at meeting a figure out of fairy tales and his growing sense of political obligation, in the end he commits the sacrilege of relegating this 900-year-old demi-deity—who has seen humans coming and going, watched their political systems and edifices rise and fall, for centuries—to just another name on the list, with his choice for "delicious" listed as "rabbit."

As Gaylen gets drawn farther into political intrigue, he begins to fulfill the Prime Minister's predictions: he has become so absorbed by his political mission and depressed by human nature that by the end of the day, as the narrator pointedly observes, "He had ridden all through a glorious afternoon, but with such a heavy heart that he had not noticed the yellow primroses starring the grass along the way or heard the sweet song of linnets in the trees."[42]

Mitt's journey into the political is similarly presented as a turning-away from his intuitive state of communion with nature and a deliberate entry into an entirely separate set of concerns and priorities. After Mitt and his family are driven off their smallholding by a vindictive taxman and forced to move to the squalid port town of Holand, the turmoil of his new human-centered life makes Mitt almost completely forget about his "perfect place." Town living leaves Mitt "shocked and shattered,"[43] and drives his perfect place out of his mind.

Jones creates an objective correlative of the "covering up" and gradual effacement of Mitt's sense of nature-divine "home" in the begrimed, effaced set of paintings in his own run-down tenement block. These had formerly belonged to a rich man, and thus have the characteristic depictions of the gods painted on its walls: flower garlands, wheat sheaves, bunches of grapes, and mythical people. As reverence for the natural deities has faded in South Dalemark, the paintings have been allowed to fade accordingly, to the point that Mitt cannot tell exactly what they depict. The mementoes of Old Ammet and Libby Beer have become too indistinct to keep Mitt within himself or to help him retain his sense of connection to nature. He reconnects with his former nature-immersed self only once, when, eluding a group of bullies, he ends up in the rich part of town and comes across a particularly fine house with relatively distinct paintings of two figures whom the reader (though not Mitt) identifies as Libby Beer and Old Ammet. These make Mitt think of his "perfect place" for the first time since leaving the countryside.

But it is also the last time for many years. Mitt soon finds himself charged by his mother with avenging his betrayed and (presumably) murdered father and "put[ting] right all the wrongs in Holand"[44] by becoming a double agent in the Marxist–type insurgency group the Free Holanders.

Parental complicity

It is both natural and significant that Mitt and Gaylen enter the world of armed class conflict not out of their own interests, but at the behest of a parent, and out of the desire to earn that parent's approval. Gaylen himself views the bickering over "delicious" absurd, and he goes back to reading his book as foster-father, the Prime Minister, declaims agitatedly about imminent civil war breaking out over it. However, when the king decides to poll his subjects and needs someone unbiased enough to trust, the Prime Minister volunteers Gaylen as being upright and honest. The more the king expresses skepticism of Gaylen's character, the more the Prime Minister insists that Gaylen is up for the job, telling the king, "I raised him myself … and he is true to the core."[45]

There is a hint of betrayal in the Prime Minister's self-identifying view of his foster-son, building on the truckling tendency demonstrated early on in his dictionary entry "Calamitous is saying no to the king"[46] and further amplified by his willingness to characterize the mission to Gaylen as an opportunity for exciting adventures, despite having already correctly predicted that the poll will trigger dissent and violence. This foreshadowing of the Prime Minister's capacity to endanger his son bears out later in the book, when he learns that the villainous character Hemlock is riding ahead fomenting insurrection against the king and the king's messenger—namely, Gaylen. Although he can easily deduce from this that his foster-son must be in very real jeopardy, the Prime Minister offers no expression of concern beyond "I am well and hope you are the same"[47] and, later, "I pray for your safety."[48]

Similarly, Mitt's mother, Milda, decides early on, entirely for her own interests, that Mitt is a "free spirit" who therefore needs no protection and who can, in fact, take on adult and even superhuman tasks to avenge his family. Frequently described as "silly," Milda badgers Mitt into joining the revolutionary Free Holanders whom she believes to have betrayed and killed her husband. She gives him a double task: first, to achieve the group's long-held aim of overthrowing the repressive Earl Hadd by assassinating him with a bomb; and second, to let himself be captured by the earl's soldiers and then inform on whichever members of the Free Holanders Mitt believes informed on his father. A careful reading of this complicated section of the book reveals that Mitt is the only one who knows that his mother is deliberately consigning her son to capture, torture, and death by hanging, and this is important to Mitt's choice at the end of the book between the social and the ecological. His new stepfather, the gunsmith Hobin, has no idea of the assassination plot or Mitt's political involvement; meanwhile, the revolutionaries in the Free Holanders fully believe that he is planning to escape.

In fact, however, Mitt has become so heavily invested in rekindling his mother's former empty-headed free-spiritedness, in taking away "the seam where her dimple used to be[, which] looked more like a puckered scar than a line"[49] with his displays of political recklessness, that he has resigned himself to death. He begins "living as if there was nothing to look forward to, beyond the moment he somehow planted a bomb under Earl Hadd."[50] When his stepfather, Hobin, tries to discipline him in aid of giving him a brighter future, Mitt takes no interest, since he knows he is never going to see adulthood.

This equation of Mitt's mother's "silliness" with a brand of genuine evil is characteristic of Jones, almost all of whose 40-plus books feature this same kind of parental (and especially maternal) neglect and cruelty. Jones herself acknowledges that many of these figures are versions of her own bohemian and self-involved mother: "[She is] in all of them, actually," Jones has said, "and she's always a villain. And, so far, she's not managed to recognize herself at all."[51] Notwithstanding this omnipresence of, in Farah Mendlesohn's words, "adults…[who] over and over …justify neglect and disparagement within a paradigm of their own deservingness,"[52] it is *Drowned Ammet* that Jones dedicates to her mother. Or rather, it is *Drowned Ammet* at the front of which she writes, "For my mother," which may be something other than a dedication.

Psychological crisis and spiritual rebirth

Just as Mitt and Gaylen have turned their backs to nature through the stronger pull of social ties—most specifically their parents, but in a larger sense the surrounding human society—they both return to their original ecocentric values after these social contracts have been violated and betrayed.

Mitt has long known of his mother's unthinkable callousness toward him, but he has survived it through enveloping himself in denial, which becomes outright dissociation by the time the assassination plot is near. It is only after he has failed in the assassination attempt and his physical body has surprised him by fleeing Earl Hadd's men of its own volition that he begins to admit the reality of Milda's actions. When he relates the assassination plot to one of the children who has been accidentally caught up in his flight, and that child asks him in bewilderment, "Your mother does know, does she, the kind of things that happen when people get arrested for your kind of thing?", his dissociative façade begins to crumble. "He felt sick and cold… . as if a load of worry had fallen on him, making his head ache and drawing his face into a strange shape."[53]

When, some time later, he learns that the ignorant and irreverent gangster who has held them hostage on board the ship is his own long-lost father, and that his father is as callously indifferent to the idea of his death as his mother was, he dissolves completely: "Mitt slid down into a heap in the corner. After a moment he wrapped his arms around his head, as if that could keep some of his misery off him. But nothing could…Mitt felt as if his whole mind was falling to pieces …There seemed nothing left of him at all."[54]

It is just at this point of seemingly complete identity loss that Mitt is restored to his former and true self. Rescued from the dungeon where his father has consigned him, he is deposited on an island by one of the indigenous Holy Islanders who recognizes Mitt's identity as part of the ecological divine: "[T]his one came on the wind's road, with a great one to guide him behind and before."[55] And here, unexpectedly, Mitt finds himself in his long-sought, long-forgotten "perfect land."

> There was a great view over the islands in the golden light, islands on one side floating green-gold in blue-grey sea, and islands on the other side blue-black against the sun, floating in silver-gold, like clouds in the sunset. Mitt, hot and breathless and miserable as he was, felt very bitter at the sight. Times out of mind, as a small boy, he had dreamed of such a place. Now he had found it, and what good had it done?[56]

But his despair at having abjured his early biophilic urges and intuitions is assuaged when the powerful being Ammet himself appears to talk with him, seriously, about where Mitt's allegiances now lie, in this time of decision. "[Y]ou will come back [to the Holy Islands]," he tells Mitt. "And when you do, it is laid down that we shall deliver these Islands into your keeping. My question to you is: 'Will you take them as a friend or as an enemy?'"[57]

Although, rather remarkably, few or none of the analyses of the Dalemark Quartet (including Farah Mendlesohn's) mention the obvious, overriding ecological import of the animistic land-gods, it is difficult to interpret Ammet's question except as posing Mitt a choice between the political (the adult, nature-repudiating and nature-exploiting social dispensation as it exists in Dalemark at that point in history) and the biocentric, biophilic one that would maintain the Holy Islands as Mitt's "perfect place." This interpretation is supported by the text immediately recasting the choice as being one between Mitt "becom[ing] the person out of his recent daydream"—a military tyrant of greater power even than Earl Hadd—and the person represented by "his own early daydreams," the ones that are full of the smell of "fresh things growing"[58] and that offer the tantalizing possibility of complete integration with the natural world. Given that Mitt has been immersed in revolutionary politics for half of his young life, abandoning his dreams of military and political domination amounts to repudiating almost all he knows of human society—and certainly all the adults he knows, including both of his parents.

Mitt resents Ammet's timing, asking why the god has to appear to him and force this choice when Mitt is shaken by his father's reappearance and the life-endangering events he has just gone through. To this, Ammet replies, "Because there was no doubt till now what you would do."[59]

Although Jones, in typical fashion, declines to explain any of the increasingly cryptic aspects of this part of the novel, one plausible reading of Ammet's statement is that the gods considered it inevitable that Mitt would be co-opted by and enveloped in the human political agenda, once that became the primary paradigm of the adults around him. (Alternatively, this hint of predestination might be part of the prophecy that we hear about in the quartet's final book, *The Crown of Dalemark*, which is never explained.)

Likewise, the reason that "it has to be now" could be that only now has he arrived in a place where the two opposing sets of priorities are starkly clear in an intellectual sense and, more importantly, at a deeper psychological level Mitt's "ultrasocial" instincts have been temporarily suspended, and the biophilic ones can take precedence. Mitt's emotional state, as he roams exhausted and traumatized on one of the Holy Islands, stripped of all his illusions and preconceptions of self, with his parents' newly realized betrayal rubbing against undeniable proof of the animistic gods' existence and their faithfulness, could represent one of the most illuminating depictions of what the choice is, and how extreme the alienation has to be in order for the biophilic to override the social.

When, out of honor for "his own early daydreams," Mitt chooses friendship with Ammet and the other inhabitants of the Holy Islands, his reward comes in the form not of any direct praise from Ammet (who merely leads his horse into the sky and disappears), but of being taught the various true names of the gods, which are too strong to be voiced "unless the speaker has right in the heart of him, and…cruel even then."[60]

For Gaylen, in *The Search for Delicious*, the spiritual crisis similarly comes at the crossroads of parental abandonment and misanthropic disillusionment. As the fight over the best food to represent the word "delicious" devolves into a genuinely life-endangering civil war between the "Crisps" and the "Squashies," Gaylen begins to acknowledge the cavalier attitude of his foster-father and the other adult authority figures who have sent him into jeopardy to fulfill their own interests. He wonders why none of the adults send for him, knowing how dangerous things are getting. He feels particularly hurt and betrayed by his foster-father, who had volunteered him for this quest even while predicting its risks: "He knows I'm all alone out here and yet there he sits, safe in the castle, tinkering with that silly dictionary. He's probably forgotten all about me."[61] The betrayal is amplified by the silliness of his grand mission to poll the kingdom about the word "delicious." Seeing the two factions' encampments, he realizes that, absurdly, the citizens of the kingdom are truly going to start a war over this petty issue.

Feelings of betrayal, abandonment, anger, and disgust temporarily snap the social bond and push Gaylen into extreme thoughts of a return to nature, in the mode of a Walden-like model of self-banishment that Laurence Buell terms "radical relinquishment."[62] He vows to take the mermaid back her lost whistle—the key to the dam that has been holding the waters back at the top of the mountain—and go live with the dwarfs or the woldweller. "They never quarrel or fight and they don't care about people any more than I do. To the devil with the poll. Hemlock can do what he likes! Why should *I* try to stop him?"[63] He even sets off in the dark to cement his new orientation by finding the mermaid, rationalizing that he is doing it not out of deep loneliness, but just because he can't sleep.

But within his fine resolution, he feels a great wave of sadness and isolation. Used and exploited by the human world, yet having no other place, he wonders where he belongs and where he can go. Although he no longer wants to ally himself with the "ninnies" who are starting a war over the definition of an adjective, he struggles to detach from society in the way he would like to, and, ultimately, finds himself unable to abandon the people he loves as they have abandoned him. A lengthy passage of internal dialogue, in italics, conveys the tension:

> *Hemlock wants to kill the King and the court!*
> *And so? It's nothing to you.*
> *But don't you see? That means the Prime Minister will be killed too.*
> *Yes, the old fool. He started it all, with his ever-lasting fried fish.*
> *Good riddance, I say.*
> *"He didn't really start it. At least, he didn't mean to. And anyway, he's my father.*
> *He isn't your father... .You're not going back to the castle anyway.*
> *You don't want to see any of them again. That's what you said.*[64]

In the short term, Gaylen is allowed to defer the decision: the villain Hemlock's threat to dam the main river and deprive the kingdom of water binds alike the fate of human and non-human, wise woldwellers and petty-minded "ninnies." The climax to *The Search for Delicious* temporarily reconciles the human with the ecological immortal, as Gaylen thwarts Hemlock by persuading the mermaid Ardis to break Hemlock's dam for all their sakes. It reconciles, too, the trivial with the essential, as the people in the kingdom in the end unanimously vote for a tasteless substance, water, as the most "delicious" food.

Anomic return to society

The two protagonists' move back into the biocentric proves as transitory as their earlier move into politics. There is, now, nowhere for them to reside free of internal conflict. Both Gaylen and Mitt return to human society and social allegiances, as, for lack of an alternative, they must. Both do it, however, with an awareness of the foolishness and destructiveness of the adult world's anti-nature values and the damage it does to their own intactness and wholeness. The two novels retract their characters from an ostensible "coming of age," yet show them to have crossed a divide the awareness of which there can be no expunging. They can move back in action, in stance, but they cannot unknow what they have learned.

Neither book presents a solution to the existential and ecological crisis it has illuminated, nor even makes much of a move to ease the anomic discomfort it has raised in its characters and, perhaps, its readers. The most *Drowned Ammet* and *The Search for Delicious* do is discard certain potential solutions as simple-minded. Their use of politics, for instance, is obviously a feint: Mitt and Gaylen learn from their stint in the world of dogma and armed conflict not which side of the class struggle to take, but rather that, in the words of Paul Virilio's famous epigram, "the ecological battle is the only one worth fighting."[65] That is, since the politics of the ecology, which sustains human life, are necessarily prerequisite to all other politics, victories in traditional political arenas are both hollow and temporary if they come at the expense of the environment. As well, both Jones and Babbitt make it clear that their conception of a Gaiaistic, holistically animated earth is not of the variety that indemnifies humans through the specious argument that, in Dorion Sagan and Lynn Margulis's summation, "we have been preceded in our massive ecocide by other life-forms."[66]

These rejections, however, offer no answer to the key question that such ecological fantasies raise. For both characters, the choice is put to them as being between the human (adult) world and the natural (childhood) one. They choose the natural world, as their moral stand and their stand for the meaningful and long-term. But both are driven by their most powerful species instinct to return to the human world.

Eventually, despite Gaylen's disgust with human society, he recognizes that the choice is not really his.

> Past his mind's eye streamed all the faces he had seen, all the kind, angry, laughing, anxious faces that had peopled the days of his great adventure. And he remembered, too, those others: the woldweller's gray cheeks, fixed into furrows like the bark of a tree; the dwarfs, impassive and calm as the mountains themselves; the wind that spoke through a hundred wayward, invisible mouths; and Ardis [the mermaid] with her eyes wet and unfathomable as the lake that glimmered before him. He leaned over and studied the dim reflection of his own face in the water… . A transient, changeable, ageable face. A people face. "And that's where I belong," he said to himself at last.[67]

After his triumph over Hemlock at the source of the river, Gaylen returns to the walled life of the palace, a decision which both his own internal dialogue and the terms of the book represent as a capitulation to the despised human "silliness." Although he promises himself that he will just stop in at the palace, say goodbye to his foster-father, and leave again, to dwell among "*the trees and the rocks and peace*," his more cynical side proves right when it predicts that the social will triumph over the ecological, sneering back at him, '*Never. You're too silly. You're a great baby. What became of all your fine anger and resolve?*'"[68]

As a married adult, Gaylen intermittently visits Ardis and the wold-weller, but human life in the kingdom otherwise proceeds as usual. Back in the palace, in the book's last pages, listening to the King and Queen squabble, Gaylen "yearn[s] again toward the unencumbered land beyond." Then he hears his foster-father calling him, and "his heart warm[s]."[69] The social has overridden the biophilic, despite the value that Gaylen, the narrator, and the book have put on the latter and the scorn with which all three have treated the former. Gaylen must live with this pull between his human social relationships and his desire to be an integral part of the living world outside the closed-off palace walls.

Mitt's fate, typically of Diana Wynne Jones's denouements, is so complex and oblique that all one can say definitively about it is that, like Gaylen's, it leaves the basic dilemma unresolved. When, near the end of the Dalemark Quartet's final book, *The Crown of Dalemark*, Mitt accepts the role of the king who might reunite the country's fractious earldoms, he must choose from various names to rule under. Each name represents a discrete aspect of the Undying self that will narrow the focus of his actions through the centuries. As himself, i.e., a kinsman or mode of Alhammitt/ Old Ammet, Mitt's urge is to live in the peace of the Holy Islands, tend "the green ways" where the gods walk, and renew the once-fertile plain known as the Shield of Oreth. Reluctantly, however, he chooses instead to rule as Amil, a name that means both "brother" and "river" in the etymology of North Dalemark proper nouns, and one of several names of the One—the River God, greatest of the Undying. The One reminds Mitt, as he is making his choice, "[Y]ou must know that I have sworn to root out [the evil Undying mage] Kankreddin from my land. If you take my name, that will be your task, too."[70]

By choosing the name Amil over Alhammitt, Mitt turns away for a second time from his ecological identity, this time literally, and pursues the more human-related agenda of keeping Kankreddin from trapping human souls before they can be swept out to sea. As King Amil the Great, he "[sees the] country through from two primitive groups of earldoms to a fully industrial society."[71] The "green roads" that have been made and maintained by the Undying are trammelled by his surrogate father-figure

Alk's railway system. With Mitt's childhood visions of the "perfect place" and a life immersed in the land now abandoned, the green roads go from having "water rilling over stones[,]...young leaves rustling overhead, casting a mix of sunspots and shadow on the racing water"[72] to being a quotidian commuter train route. The skies are filled first with coal smog and then with car emissions and pollution that make country-bred Maewen cough.[73]

Whether Mitt feels this as a loss we never know, as this is almost the end of his story. Because *Drowned Ammet* ends with Mitt apparently unknowing that he is an immortal, and the end of *The Crown of Dalemark* returns to the point of view of Maewen, the girl from the 20th century, Mitt's emotions about his relationships with mortal human society and with nature, along with many other key elements of coherence in the quartet, fall into what Farah Mendlesohn terms "the interstices of knowledge and assumption"[74] that characterize Jones's approach to story creation. However, we can take a clue from another immortal protagonist in Jones's oeuvre, Jamie of *The Homeward Bounders* (1981). Jamie, who accidentally becomes a player in a set of multiversal war games being played by powerful demons, in the end volunteers to keep walking "the bounds" of a set of parallel worlds so that their other inhabitants, free from this duty, can dwell in them and thus keep them "real places" rather than memories that can be manipulated by the demons. Jamie's refrain throughout the book is, "You wouldn't believe how lonely you get."[75]

Whether Mitt's loneliness is eased by the presence of other immortals and deities, along with his marriages to mortal women within their lifespans and his quarter-millennium–long dedication to Maewen, is left unclear. However, Jones' descriptions of the green roads and undesecrated land are sufficiently rhapsodic and evocative to indicate that she means the reader to view Mitt's choice as a loss, and to wish that he had made a different decision when he was asked to embody either an aspect of the Earth Shaker or an aspect of Amil the River God. In this sometimes-archetypal return-of-the-king story, the usually triumphant crowning is marked with melancholy. The reader understands that by choosing the human and social, Mitt has to a large extent abjured the natural. Even writing from an approach of psychological development, fantasy theorist and Jones scholar Farah Mendlesohn sees this conflict: "[Jones's] children [in the Dalemark Quartet]...each has been corrupted by his or her own context, and until those contexts change they are unquestioning of their own culture's assumptions. The more drastic the shift in their experience, the more they are antagonists within the tale, increasingly resisting their culturally 'intended' narrative."[76]

Thus, in what could be seen as an unusual variation on the "home-away-home" pattern, both novels have their characters cross the divide between childhood and adult values, priorities, and affinities, and then return again. Mitt and Gaylen come of age not by finally repudiating their original ecocentric outlook, but by reclaiming it and then living with the internal tension between their two sets of needs. Maturity for both protagonists entails a threefold understanding: first, that environmentally irreverent or ecologically predacious attitudes are fatally short-sighted; secondly, that rejecting these attitudes means alienating themselves from the main of human society; and thirdly, that removing themselves from the main of human society is not a viable option even knowing what they know and feeling as they feel.

In Mitt and Gaylen's journey and their arrival at a place of irreconcilable tension, the psychological path of the Western child reader is metanarratively laid out. The innate fascination, reverence, and affinity for the complex and life-giving natural world will give way to a focus on human concerns, but once that social world is understood and integrated, it forces a choice between that focus and the other. Even when the triviality and self-defeatism of a wholly anthropocentric, nature-rejecting value set are clear, there is no way to go back, because to do so means to separate oneself from the majority of human society, as untenable an option immediately as that of separating from the biological matrix is ultimately.

The dilemma unresolved

That Mitt and Gaylen reach the end of their stories still in a state of conflict, with their renewed orientation toward nature serving only to make them more anomic within the human-centered status quo and more aware of their anomie, offers one suggestion as to why the biophilia-fostering role of children's literature remains an open secret rather than a topic of direct discussion. There is no easy solution to the issues of modern Western society's alienation from and degradation of the natural world; it seems likely that there is no solution at all. When *Drowned Ammet* and *The Search for Delicious* take the unusual step of depicting not only the stereotypically happy biocentric world of childhood but its nature-repudiating adult opposite, they touch on a number of uncomfortable truths and draw discomfiting attention to the ways in which the abundance of biocentric content in children's literature constitutes a certain tradition of denial and responsibility-shifting on the part of adult Western society.

Bob Henderson, Merle Kennedy, and Chuck Chamberlin, exploring the fairness of inserting these issues overtly into children's literature, propose that doing this might "wrongly burden youth while letting adults off the hook."[77] They write in regard to probably the most famous instance in children's literature of open environmental proselytizing, Dr Seuss's *The Lorax*,

> Is the [book's antagonist, the] Once-ler, as an adult, simply abdicating responsibility to the younger generation? The Once-ler's anguish for what he has done ... leads to no personal volition other than to pass on the responsibility of righting his wrong to a young boy. Is this a common practice of adults, such that children are burdened? ... The Once-ler exemplifies a standard adult practice here. [I]ntergenerational equity for the resources ... is left for the next generation to resolve, with no guidance from the generation that used the resources. This unhappy ending is a common response by adults like the Once-ler.[78]

The Search for Delicious and *Drowned Ammet* suggest an even greater abdication of responsibility, in that they highlight the deleterious effects of human alienation from nature not only on nature, but on humans. Both these novels (*Drowned Ammet* perhaps more than *The Search for Delicious*) go beyond a mere conservation ethic to take fairly seriously what has become one of the main cries of sociobiology and evolutionary psychology: as an animal species that evolved within the matrix of life and non-living natural elements, people genuinely need immersion in the non-human living world, and they suffer concrete and lasting damage without it. Thus, while it as true of these two novels as it is of *The Lorax* that "extending its logic would lead to a reexamination of American lifestyles,"[79] it is also true that examining its implications would probably lead to the conclusion that the modern Western state of urbanized Judeo-Christian techno-industrialism is contributing significantly to children's and adolescents' unhappiness and disaffection, and perhaps even making these pathologies inevitable.

That said, I believe it is important to point out that the two novels propose—and therefore may generate in their readers—quite different senses of how deep and insoluble the problem is, in terms of both environmental degradation and spiritual-psychological harm. The discrepancy follows quite closely what Louise Westling names as "significant national differences" between the American and British ecocritical traditions. The Cooperian *Search for Delicious* exemplifies Westling's typification of "[American] nature writing [as]... providing unmediated access to an essential non-human world,"[80] while the more pastoral *Drowned Ammet* is in many ways typical of the British mode, which "emerged from within an older literary tradition concerned with long-domesticated and densely populated landscapes and, for the past 200 years, with problems of industrialization and urban space."[81] Simply put, the British are in nothing like the state of despair that Americans (and to a large extent Canadians) are. The reasons for this are manifold, and arise from a complex combination of history, geography, and culture, but the difference is discernible and important in its implications.

Jones's book is what Kate Soper would term "monist"[82]: it presents humans as being necessarily and obviously part of the natural matrix, and presents the main question therefore as being whether we acknowledge and live by this or deny and destroy it. The consequences of doing the latter are mainly depicted in terms of the loss to us—spiritually, existentially, practically, rationalistically—and the option is presented as a still-viable choice mostly involving a return to earlier (and not-long-dormant) attitudes toward nature and a relatively recent agrarian lifestyle. In the Dalemark Quartet overall, and *Drowned Ammet* in particular, pre-Christian, ecocentric spiritual/religious beliefs (in this case, a version of the animistic polytheism of the Celtic tribes) are the bedrock upon which new, nature-rejecting attitudes are thinly and incompletely grafted. The old beliefs continue to percolate up, and they are not only easily triggered but often unknowingly acted upon.

In keeping with these depictions, the novel (and quartet) conveys a relative lack of extremism, a lack of despair, and a space to assign people value on par with the rest of the ecology. A passage in *The Crown of Dalemark* in which Mitt is trying to decide whether to accept the crown of Dalemark, very near the quartet's end, signifies this attitude: "Mitt knew it was a real choice, even if Maewen [the girl from the 20th century, who already knows what happened in Mitt's time] told him which way he chose."[83] The decisions of the past, even ones that would seem to determine the future, do not necessarily preclude different choices in the present.

As well, in keeping with the pastoral mode, a large part of the solution is presented as a matter of geographical rather than chronological shift. Living within a degraded natural matrix, as portrayed in the fate of the Shield of Oreth, can be rectified either by moving to a different place or "restoring" the degraded space to its pastoral condition, both actions easily achievable in a short space of time. The key idea here is that of return: to a way of living, an attitude toward nature, and a condition of the land that lies within recent historical memory. The reader of the Dalemark Quartet may come away with a sense of mourning for the unspoiled pre-Industrial landscape, but the solutions to this are offered within the quartet itself. For instance, in *The Crown of Dalemark*, the country-raised modern-day girl Maewen finds the big city noisy, polluted, and generally "distressing," and she particularly deplores the absence of trees[84]; however, she knows that her residence there is only temporary and that she can return to her life surrounded by horses and fields as soon as her visit with her father ends. In a larger ecological context, the Shield of Oreth does get "restored" to fertility, and even if railroads and the city of Kernsburgh have subsequently been erected on it, we know that in a span of generations the earth can and will reclaim both, for nature in the quartet is always in the process of returning to itself.

The Search for Delicious, notwithstanding the critics' labeling of it as "delightful," "subtly teasing" and "winning," offers a much bleaker, irresoluble, and misanthropic vision of humanity's turn away from biophilia than does *Drowned Ammet*. Species self-loathing and despair seem to be the only appropriate emotional responses to a problem that has an uncaused commencement and is as intractable as it is disastrous. In *The Search for Delicious*, there is nowhere for Gaylen or the humanity he represents to return to. Humans come from nowhere, destroying and quarreling as they go, long after the arrival of the other creatures, who are part of the natural immanence. They have no place at the time of the events in the novel—and these people, European-descended, nature-scorning, "civilized" people, never have.

Although when Gaylen sees fish dying in the drying-up river, near the end of the book, he realizes that "the people of the kingdom were facing the same grim possibility,"[85] his recognition of mutual dependency on the water sources does not lead to the suggestion that humans might learn and change, nor reverse the forfeiture of a place in the ecosystem (and the narrator's esteem) born of their abiding foolishness and lust for power. No alternative is offered, either in the future or the past, except, as I have pointed out, that of "radical relinquishment" and hermitude. With Gaylen's self-betraying return to the human fold, the book shows that the fundamental need for society puts this option outside the capacity of even the strongest-willed person.

The solution-less conclusion that Babbitt offers, in juxtaposition with the gravity of the problem that she outlines, is the kind of Calvinist vision that dominates American post-apocalyptic ecological fiction, such as Don DeLillo's *White Noise* and Carl Hiaasen's *Sick Puppy*. Briton Richard Kerridge writes with some puzzlement about the "ironic neutrality"[86] in books featuring "the aftermath of an ecological collapse so total as to preclude any continuance of industrial capitalism,"[87] non-dialectical models where "the apocalyptic allows for no return" and "therefore no mediation between the [pre-industrial] world and our own."[88] *The Search for Delicious*, and to a somewhat lesser degree its dizygotic twin *Tuck Everlasting*, fall under Kerridge's admonishing heuristic of "novels [which] achieve their most powerful effects by setting their calamities beyond the reach of remedy," that "strikingly" lack any belief "that a pragmatic, incremental environmentalism, allied to other social movements, might begin to turn the tide," and which therefore "take failure for granted."[89]

In the end, both *Drowned Ammet* and *The Search for Delicious* ask, but do not answer, the question upon which much children's literature stays silent: in Charles Butler's words, "What is the optimal state, the one in which humanity and nature may live in some degree of harmony?"[90] They make evident that if, as China Mieville claims, "the best fantasy" is characterized by a "rejection of consolation,"[91] the same is also true of the best ecological writing. The inescapable import of this is that, at least for the children's literature that is themed around nature and animals, the "happy endings" ascribed to the genre might be more accurately described as the avoidance of an ending. From an ecocritical view, indeed, it could be said that the only children's books with a genuinely happy ending are those that involve themselves solely with social and relationship themes and do not draw in the ecological context.

Notes

1 Greg Garrard, *Ecocriticism* (Routledge, 2004), 19.
2 Garrard, *Ecocriticism*, 19.
3 Garrard, *Ecocriticism*, 21.
4 Garrard, *Ecocriticism*, 20.
5 E. O. Wilson, *Biophilia* (Harvard University Press, 1984), 85.
6 Arne Naess, "The Deep Ecological Movement," in *Deep Ecology for the 21st Century: Readings on the Philosophy and Practice of the New Environmentalism*, ed. Frederik A. Kaufman (Shambhala, 1995), 68.
7 Colin Manlove, *The Fantasy Literature of England* (Macmillan, 1999), 83.
8 Diana Wynne Jones, *The Crown of Dalemark* (Oxford University Press, 2003), 263.
9 Natalie Babbitt, *The Search for Delicious* (Farrar, Straus and Giroux, 1989), 6.
10 Babbitt, *Search for Delicious*, 7.
11 Wilson, *Biophilia*, 41.
12 Wilson, *Biophilia*, 43.
13 Diana Wynne Jones, *Drowned Ammet* (HarperCollins, [1977] 2016), 14–15.
14 Jones, *Drowned Ammet*, 18.
15 Jones, *Drowned Ammet*, 23.
16 Babbitt, *Search for Delicious*, 13.
17 Babbitt, *Search for Delicious*, 61.
18 Babbitt, *Search for Delicious*, 27.
19 Babbitt, *Search for Delicious*, 10.
20 Babbitt, *Search for Delicious*, 36–37.
21 Babbitt, *Search for Delicious*, 37.
22 Babbitt, *Search for Delicious*, 37.
23 Jones, *Drowned Ammet*, 12.
24 Jones, *Drowned Ammet*, 258.
25 Babbitt, *Search for Delicious*, 45.
26 Scott Slovic, "Epistemology and Politics in American Nature Writing: Embedded Rhetoric and Discrete Rhetoric," in *Green Culture: Environmental Rhetoric in Contemporary America*, ed. Carl Herndl (University of Wisconsin Press, 1996), 82.
27 Slovic, "Epistemology and Politics," 85.
28 Babbitt, *Search for Delicious*, 5.
29 Slovic, "Epistemology and Politics," 83.
30 Nick O'Connell. "Writer's Interview with Barry Lopez," *The Writer's Workshop*, https://www.thewritersworkshop.net/about-us/writers-workshops-nicholas-oconnell-interview-barry-lopez/. Qtd in Scott Slovic, "Epistemology," 84.
31 Babbitt, *Search for Delicious*, 41.

32 Farah Mendlesohn, *Rhetorics of Fantasy* (Wesleyan University Press, 2008), 1.
33 Babbitt, *Search for Delicious*, 42.
34 Babbitt, *Search for Delicious*, 5.
35 Babbitt, *Search for Delicious*, 8.
36 "How Do Books Get Banned?" *First Amendment Museum*, accessed January 21, 2026. firstamendmentmuseum.org.
37 Perry Nodelman, *The Hidden Adult: Defining Children's Literature* (Johns Hopkins University Press, 2008), 150.
38 Nodelman, *Hidden Adult*, 121.
39 Jacques Lacan, qtd in Coats, *Looking Glasses and Neverlands*, 4.
40 Bruno Bettelheim, *The Uses of Enchantment: The Meaning and Importance of Fairy Tales* (Knopf, 1991), 46.
41 Wilson, *Biophilia*, 118.
42 Babbitt, *Search for Delicious*, 72.
43 Jones, *Drowned Ammet*, 29.
44 Jones, *Drowned Ammet*, 45.
45 Babbitt, *Search for Delicious*, 25.
46 Babbitt, *Search for Delicious*, 14.
47 Babbitt, *Search for Delicious*, 40.
48 Babbitt, *Search for Delicious*, 140.
49 Jones, *Drowned Ammet*, 38.
50 Jones, *Drowned Ammet*, 77.
51 Farah Mendlesohn, *Diana Wynne Jones: Children's Literature and the Fantastic Tradition* (Routledge, 2005), xxiv.
52 Mendlesohn, *Diana Wynne Jones*, xxiv.
53 Jones, *Drowned Ammet*, 217.
54 Jones, *Drowned Ammet*, 323.
55 Jones, *Drowned Ammet*, 325.
56 Jones, *Drowned Ammet*, 331–32.
57 Jones, *Drowned Ammet*, 334.
58 Jones, *Drowned Ammet*, 15.
59 Jones, *Drowned Ammet*, 334.
60 Jones, *Drowned Ammet*, 339.
61 Babbitt, *Search for Delicious*, 116.
62 Lawrence Buell, *The Environmental Imagination: Thoreau, Nature Writing, and the Formation of American Culture* (Harvard University Press, 1995), 144.
63 Babbitt, *Search for Delicious*, 117.
64 Babbitt, *Search for Delicious*, 130.
65 Paul Virilio, *Popular Defense & Ecological Struggles* (Semiotext(e), 1990), qtd in *The Green Studies Reader*, 5.
66 Dorion Sagan and Lynn Margulis, "God, Gaia, and Biophilia," in *The Biophilia Hypothesis*, eds. Stephen R. Kellert and Edward O. Wilson (Island Press, 1993), 349.
67 Babbitt, *Search for Delicious*, 148.
68 Babbitt, *Search for Delicious*, 131.
69 Babbitt, *Search for Delicious*, 160.
70 Jones, *The Crown of Dalemark*, 283.
71 Jones, *Crown*, 59.
72 Jones, *Crown*, 51.
73 Jones, *Crown*, 65.
74 Mendlesohn, *Diana Wynne Jones*, 118.
75 Diana Wynne Jones, *The Homeward Bounders* (HarperCollins, 2000), 36, 267.

76 Mendlesohn, *Diana Wynne Jones*, 123.
77 Henderson, Kennedy, and Chamberlin, "A Pedagogical Response," 139.
78 Henderson, et al., "A Pedagogical Response," 139.
79 Henderson, et al., "A Pedagogical Response," 142.
80 Westling, "Introduction," 1–2.
81 Westling, "Introduction," 3.
82 Soper, *What Is Nature?*, 49.
83 Jones, *Crown*, 283.
84 Jones, *Crown*, 65.
85 Babbitt, *Search for Delicious*, 125.
86 Richard Kerridge, "Narratives of Resignation: Environmentalism in Recent Fiction," in *The Environmental Tradition in English Literature*, ed. John Parham (Ashgate, 2002), 97.
87 Kerridge, "Narratives of Resignation," 96.
88 Kerridge, "Narratives of Resignation," 97.
89 Kerridge, "Narratives of Resignation," 99.
90 Charles Butler, *Four British Fantasists: Place and Culture in the Children's Fantasies of Penelope Lively, Alan Garner, Diana Wynne Jones, and Susan Cooper* (Scarecrow Press, 2006), 132.
91 Joe Sutliff Sanders, "Reinventing Subjectivity: China Mieville's *Un Lun Dun* and the Child Reader," *Extrapolation* 50, no. 2 (2009): 293.

Bibliography

Babbitt, Natalie. *The Search for Delicious*. 1969. Farrar, Straus and Giroux, 1989.
Babbitt, Natalie. *Tuck Everlasting*. 1975. Farrar, Straus and Giroux, 1991.
Bettelheim, Bruno. *The Uses of Enchantment: The Meaning and Importance of Fairy Tales*. 1976. Knopf, 1991.
Buell, Laurence. *The Environmental Imagination: Thoreau, Nature Writing, and the Formation of American Culture*. Harvard University Press, 1995.
Butler, Charles. *Four British Fantasists: Place and Culture in the Children's Fantasies of Penelope Lively, Alan Garner, Diana Wynne Jones, and Susan Cooper*. Scarecrow Press, 2006.
Coats, Karen. *Looking Glasses and Neverlands: Lacan, Desire and Subjectivity in Children's Literature*. University of Iowa Press, 2004.
Dobrin, Sidney I., and Kenneth B. Kidd, eds. *Wild Things: Children's Culture and Ecocriticism*. Wayne State University Press, 2004.
Garrard, Greg. *Ecocriticism* (Routledge, 2004), 19.
Henderson, Bob, Merle Kennedy, and Chuck Chamberlin. "Pedagogical Response to *The Lorax*." In *Wild Things: Children's Culture and Ecocriticism*, edited by Sidney I. Dobrin and Kenneth Kidd. Wayne State University Press, 2004.
"How Do Books Get Banned?" *First Amendment Museum*. Accessed January 21, 2026. firstamendmentmuseum.org.
Jones, Diana Wynne. *The Crown of Dalemark*. 1993. Oxford University Press, 2003.
Jones, Diana Wynne. *Drowned Ammet*. 1977. HarperCollins, 2016.
Jones, Diana Wynne. *The Homeward Bounders*. 1981. Collins, 2000.
Jones, Diana Wynne. *The Spellcoats*. Macmillan, 1979.
Kellert, Stephen R., and Edward O. Wilson, eds. *The Biophilia Hypothesis*. Island Press, 1993.

Kerridge, Richard. "Narratives of Resignation: Environmentalism in Recent Fiction." In *The Environmental Tradition in English Literature*, edited by John Parham. Ashgate, 2002.
Manlove, Colin. *The Fantasy Literature of England*. Macmillan, 1999.
Mendlesohn, Farah. *Diana Wynne Jones: Children's Literature and the Fantastic Tradition*. Routledge, 2005.
Mendlesohn, Farah. *Rhetorics of Fantasy*. Wesleyan University Press, 2008.
Naess, Arne. "The Deep Ecological Movement." In *Deep Ecology for the 21st Century: Readings on the Philosophy and Practice of the New Environmentalism*, edited by George Sessions. Shambhala, 1995.
Nodelman, Perry. *The Hidden Adult*. Johns Hopkins University Press, 2008.
Richerson, Peter J., and Robert Boyd. "The Evolution of Human Ultra-Sociality." Version 2.01, 1997. sscnet.ucla.edu/anthro/faculty/boyd/ultra.pdf. Online PDF.
Sagan, Dorion, and Lynn Margulis. "God, Gaia, and Biophilia." In *The Biophilia Hypothesis*, edited by Stephen R. Kellert and Edward O. Wilson. Island Press, 1993.
Sanders, Joe Sutliff. "Reinventing Subjectivity: China Mieville's *Un Lun Dun* and the Child Reader." *Extrapolation* 50, no. 2 (2009): 293–307.
Sessions, George, ed. *Deep Ecology for the 21st Century: Readings on the Philosophy and Practice of the New Environmentalism*. Shambhala, 1995.
Slovic, Scott. "Epistemology and Politics in American Nature Writing: Embedded Rhetoric and Discrete Rhetoric." In *Green Culture: Environmental Rhetoric in Contemporary America*, edited by Carl Herndl. University of Wisconsin Press, 1996.
Soper, Kate. *What Is Nature? Culture, Politics and the non-Human*. Blackwell, 1995.
Stierstorfer, Klaus, ed. *Beyond Postmodernism: Reassessments in Literature, Theory and Culture*. Walter de Gruyter, 2003.
Westling, Louise. "Introduction." In *The Environmental Tradition in English Literature*, edited by John Parham. Ashgate, 2002.
Wilson, Edward O. *Biophilia*. Harvard University Press, 1984.

3 "The belief in it a blunder, the hope of it a sin"

Honor as anti-immortalism and species self-limitation in *The Farthest Shore* and *Tuck Everlasting*

Ursula K. Le Guin's *The Farthest Shore* (1972) and Natalie Babbitt's *Tuck Everlasting* (1975) hold particular importance to my thesis in presenting what are perhaps the most forceful and explicit anti-immortalist arguments to date in children's literature, or perhaps in any literature. Rather than being a theme or implication, the argument against the pursuit of eternal physical life is their narrative subject, their thesis, and their sole focus. Further, they base their arguments on a complex spiritual-ecological ontology that bears many hallmarks of Deep Ecology. In the process, the books sharply delineate the inverse relationship between biocentric values and the ambition of avoiding death and achieving physical immortality.

The Farthest Shore details the quest of Earthsea's powerful Archmage Ged and his companion, the young prince Arren, to find and defeat the mage who has opened a gap in the wall between life and death. This undead mage, Cob, has begun "selling" a return path from the land of death to those frightened enough of death to give him their true names as the price.

Tuck Everlasting starts with the Tucks, an American pioneer family who accidentally became immortal by drinking from a spring of elixir vitae, trying to persuade 10-year-old Winnie Foster not to drink from the spring of elixir vitae. It then moves on to the protagonists' fight against the immortalist villain who is scheming to sell the water.

Michael R. Collings' summary of *The Farthest Shore* applies equally to *Tuck Everlasting*: "[The] indictment of 'unnatural' immortality goes beyond merely arguing that those who receive this dubious boon somehow lessen themselves as human, [and] concentrates on the consequences for an entire world of such a bending of the natural order."[1] The mentor characters in each novel—Ged, in the *Farthest Shore*, and the three oldest members of the Tuck family, in *Tuck Everlasting*—spend literally pages impressing upon their young protégés what will happen if immortalist thinking takes hold: nothing less than the undoing of the world as we know it. Violating nature's ultimate rule in these books does not merely disqualify a person from a position of high leadership, but begins a process

DOI: 10.4324/9781003784210-4

of existential annihilation. Ged tells Arren that in creating a two-way passage between life and death, the mage Cob has begun to disturb the great Equilibrium at its roots, so that both things and people are losing their essential identities, roles, and capacities that allow the world to exist. Pa, Ma, and Miles Tuck tell Winnie that to be taken off the great wheel of life is to cease to be, to lose all purpose and meaning, "like rocks beside the road."[2]

The two novels can be seen as anti-technological, and more specifically as responding to contemporaneous social issues and scientific advances (or the promise of them) that I will enumerate briefly below. Less obviously, in their refusal to assign humanity a position above the web of nature or beyond the reach of its consequences, they also reject the human exceptionalism upon which Judeo-Christianity rests.

Tuck Everlasting presents a web-of-life view that comes even closer to Deep Ecology than does Babbitt's earlier *The Search for Delicious*, while *The Farthest Shore*, like the rest of Le Guin's Earthsea Trilogy, argues for so thoroughgoing an essentialism that it might be thought of as Deep Animism. In short, both books openly foster worldviews that by any measure challenge the Christian-based, post-Industrial precepts that supposedly make up the norms of the Western outlook. And though they weave their theses deftly and in many ways subtly, each book contains several anti-immortalist monologues of remarkable length and rhetorical vigor.

Yet, despite their overtly agendistic natures, these books are considered classics of children's literature and are perennial backlist bestsellers: *Tuck Everlasting* has twice been made into a major feature film, and both books have been translated widely and kept continually in print for the past five decades. That two such openly pro-ecological, anti-exceptionalist novels can be not only accepted but embraced points to the extent to which a Western child reader's education in "ecological literacy" is a radical and subversive one, and supports my contention that the publishing tradition of children's literature acts as a powerful agent for biocentricity.

A new urgency for honor

The Farthest Shore and *Tuck Everlasting* mark a newly prominent focus on the character trait of honor as humanity's best answer to the scourge of immortalist thinking. Here I must pause briefly in my analysis to outline the conception of honor that I use in this chapter, since it is (to quote the White Knight) something of "my own invention." While the dictionary defines the noun "honor" merely as possessing "a strong sense of what is right; keen moral judgment," in its applied sense it means declining to act on a power advantage—in other words, not doing something simply because we can, or, conversely, doing something even though we don't have to. Applied by human populations to their behaviors regarding the natural environments, honor by this definition amounts to an inherent tendency toward species self-limitation and against ecologically destructive behaviors.

Most environmental and ecological thinkers grant humanity no such inherent tendency. They assume that the human species possesses no innate mechanism of self-limitation to offset what E. O. Wilson terms "the audaciously destructive tendencies of our species."[3] Instead, they write from the perhaps unnecessarily misanthropic premise that humans, having developed technologies to overcome the natural checks imposed by the environment and having thereby essentially adapted our way into disequilibrium, are now running unfettered, misguided by our myopic sense of "biological time" into destructive behavior that serves only our short-term interests.

The field of evolutionary ethics contributes by omission to this concept. Granted, the descriptive branch of the field, which examines how ethics may have arisen as ways to improve group fitness and genetic competitiveness, marks an improvement over the normative branch, which involves itself almost wholly in trying to reverse-engineer a Christian morality out of Darwinian precepts. Nonetheless, both forms of evolutionary ethics focus on the intra-human functions of "virtues" such as altruism and honesty. In doing so, they reinforce the notion that we have evolved no biocultural mechanism, nor even a predisposition, that would facilitate a longer-term, extra-species awareness of self-restraint as conferring a survival advantage within the larger biological matrix.

To me, this idea makes no particular sense. Why, given that we have evolved to possess the capacities of self-awareness, foresight, and an inherent understanding of our dependence on natural systems, would we have so entirely failed to integrate these capacities as to remain devoid of an intuitive sense of "ecological time," and thus be unable to respond to multigenerational threats to our environment, as Wilson and others contend, except "through an unusual amount of education and reflective thought"?[4] Likewise, why would the moral sensibilities of various human populations, which according to evolutionary ethicists arose to "[give] certain groups a selective advantage" and which "[can] be validated by reference to their adaptive value," to use Paul Lawrence Farber's summation,[5] have evolved in isolation from the natural systems upon which we depend and without which no amount of genetic, group, or species fitness is of any avail?

By any commonsense extension of sociobiological premises, an evolved and innate ecological ethic must exist. I would argue that in fact it does exist, and that one of its chief mechanisms is the "virtue" of honor. As I say above, I use the noun "honor" to mean, in its most common application, refraining from acting on a power advantage. One of the most traditional contexts for this is what Ted Westhusing terms "Warrior honor":

"a certain constancy, harmony and refinement of the natural virtues of greatness of mind and extended benevolence. Both virtues for the warrior are deeply rooted in and expressive of a common life for which he is prepared to die."[6]

This kind of honor is, needless to say, a common trope in both children's and adult fantasy. In a number of children's texts, however, this ultimate show of honor—the willingness to die in battle—is highlighted by its subtler obverse: the rejection of a chance to live forever. In such canonical children's fantasy as Tolkien's Lord of the Rings and Lloyd Alexander's Prydain Chronicles, it is not enough that the heroes risk their lives in battle for "a common life" toward which they feel self-sacrificing allegiance. In the end, when the fighting is done, they must reject the temptation of eternal life as the acid test of their honor. Only when they have done this are they given the right to lead the whole of society found within the books, in both cases by being made High King. Given the popularity and high sales volumes of these books, we can surmise that a good proportion of the Western child readership is receiving an osmotic education in the concept of self-limitation. Thus, while E. O. Wilson may be right in believing that we must be educated into a sense of longer-range vision and "a high premium on posterity," he may be wrong in believing that such an education is "unusual."[7]

Tolkien and Alexander's books treat the honorable rejection of immortality as symbolic—symbolic, granted, of an important premium on self-restraint, but still in the register of high fantasy, where the immortality on offer is portrayed in a highly subjunctive manner and, in fact, not as malign in itself. In *The High King*, Taran turns down passage to the Summer Country of the Sons of Don, with its eternal life and perfect happiness—but although this is the requisite action to his being named High King and leading Prydain into a golden age, other main protagonist characters do cross to the vaguely described Summer Country without authorial indictment. In *Return of the King*, Aragorn overcomes his race's long and disastrous history of striving for immortality when he chooses to enter permanent sleep before he has to, but he and an array of other mortal species co-exist side by side with the immortal elves, who are among the most favorably depicted characters in Tolkien's subcreation.

So although we can infer that in these books an honorable rejection of immortality represents a mindset and character that is crucial for the arch-rulers of a wide group of cultures (Aragorn and Taran both being high kings rather than merely regional kings), the idea of human physical immortality is not demonized, because it is not addressed with any seriousness.

The Farthest Shore and *Tuck Everlasting* treat immortality, and the honor and courage to reject it, in a wholly different register. In its simplest terms, the difference is that these two books take the idea of human physical immortality seriously—as not just a concept, but a prospect; not just a symbol, but a potentiality. In this, Le Guin's trilogy and *Tuck Everlasting* are both products of their era and representatives of the fantasy genre. At the time of their publication in the late 1960s and early 1970s, as Merritt Abrash observes, medical advances and breakthroughs in the nascent field of biotechnology, such as the discovery of the so-called "hibernation induction trigger" in the brain's hypothalamus,[8] meant that "the theme of bodily immortality, once imaginable only as a manifestation of ancient legend or exotic travel fantasy, [had come] to be treated within a framework of scientific plausibility—well, let us say pseudo-possibility—and correspondingly more realistic psychological and institutional consequences."[9] This meant that in treating the subject of immortality, both fantasy and science fiction writers now had to move at least partly out of their customary registers of what Samuel Delany calls "subjunctivity"—fantasy the deeply subjunctive register of "this could not have happened," science fiction the shallower one of "this has not happened"—and into the realm of mimesis, of realistic fiction's "this might happen."[10]

Science fiction responded with enthusiasm, fantasy with dismay, and the age-old generic battle of science versus prescience was rejoined. Alan Harrington's 1969 pro-immortality manifesto *The Immortalist* expresses the science-fiction/science-fact position in terms that reflect the euphoria of the era and presage the fear-driven anti-death campaign of today's biotechnology industry:

> Death is an imposition on the human race, and no longer acceptable. Man has all but lost his ability to accommodate himself to personal extinction; he must now proceed physically to overcome it... . [W]hy should death itself, "the Last Enemy," be considered sacred and beyond conquest? ... We must drive away the gods of doubt and self-punishment. Our new faith must accept as gospel that salvation belongs to medical engineering and nothing else; that man's fate depends first on the proper management of his technical proficiency; that we can only engineer our freedom from death, not pray for it; that our messiahs will be wearing white coats, not in asylums but in chemical and biological laboratories.[11]

Judging by the overtly anti-immortalist focus of *The Farthest Shore* and *Tuck Everlasting*, Ursula K. Le Guin and Natalie Babbitt took alarm—if not at the prospect of unnatural life extension itself, then at least with the biophobic, anti-ecological immortalist thinking that its newly "pseudo-possible" status seemed to encourage among the Western populace. Both authors give pride of place to honor as the key heroic characteristic—and, specifically, the honor to reject immortalist thinking. The child who reads either book will receive a persistent and explicit education in the need for self-conquering and self-limitation.

Moreover, the authors use the trait of honor-as-self-restraint metonymically to stand in for species self-restraint. In so doing, they anticipate the second wave of evolutionary ethics, specifically the "biophilia" hypothesis that E. O. Wilson would develop in the 1980s. This proposed, among other things, that apparently self-sacrificing human social and cultural behaviors, which social scientists recognized as belonging to "enlightened self-interest," in the long term include interactions with and attitudes toward the natural world.

Power restrained

The Farthest Shore presents a lengthy answer to a crucial question about honor that Ged asks himself in the Earthsea cycle's first book, *A Wizard of Earthsea*. A willful young apprentice, Ged is then still at a point of immaturity when "all his pleasure in the art-magic [is] childlike, the power it [gives] him over bird and beast."[12] Having been mistaught early by a village witch who "knew nothing of the Balance and the Pattern which the true wizard knows and serves, and which keep him from using his spells unless real need demands,"[13] Ged endures with ill grace his new master Ogion's continual self-restraint. He reaches a pinnacle of resentment when Ogion will not even use the simplest of weather-working spells to keep the two of them dry: "Ogion let the rain fall where it would. He found a thick fir-tree and lay down beneath it. Ged crouched among the dripping bushes wet and sullen, and wondered what was the good of having power if you were too wise to use it."[14]

By the time of *The Farthest Shore*, Ged, now Archmage, is deeply schooled in the answer to that question: it is (as Caspar said of his royal blood in *Amahl and the Night Visitors*) "No use." As he tells his royal protégé Arren, the long education that a prospective wizard undergoes on the Isle of Roke is, more than anything, a steady indoctrination into a continual forgoing of the use of power. "The first lesson on Roke, and the last, is *Do what is needful*. And no more!"[15]

Arren's main activity in the book, and in fact his main *raison d'etre* altogether, is to receive, register, and reflect the lessons in honor that Ged has learned in his lifetime. In terms of its theme and thesis, *The Farthest Shore* could be seen as a reiteration of *A Wizard of Earthsea* with the Jungian overtones eliminated and the character-quest for honor isolated, amplified, and set toward one target: that of countering immortalism. Ged says as much, both before their journey, when he explains to the other mages that he has finally found a worthy companion, and near its end, when he says to the sleeping Arren, "I have found none to follow in my way. None but thee."[16]

Le Guin has Arren filter much of his early perceptions of the great mages' home, the Isle of Roke, through the lens of self-restraint. Arren notes, for instance, that these mages, the nine most powerful humans in Earthsea, live lives of monastic simplicity. For Prince Arren, who has grown up knowing wealthy traders and aristocrats and is used to their self-interested bargaining and quarreling, this calm self-divestment of power is new and startling. The mages are quiet men with simple tastes and quiet faces. "If they sought something, it was not for themselves. Yet they were men of great power: that, too, Arren recognized."[17] He responds in similar fashion to the young Ged, if with more princely grace and less arrogance, when the Archmage emulates his master Ogion and refuses to sail his small boat by summoning a wind, as a mere weatherworker could and would have done. In fact, after several days as Ged's companion, Arren realizes that the Archmage has thus far performed no magic at all.

In their relationship also, Ged demonstrates honorable restraint in refraining from using his power over Arren. He knows Arren's true name, Lebannen, which means "rowan" in Arren's native Enladian language, but does not use the power this gives him, or even refer to it except in a casual mention of rowan trees meant to alert Arren (honorably) to the fact that he knows it. He impresses upon Arren that choosing to journey with him, rather than returning to his royal comforts after completing his task as an emissary, is the choice between safety and peril. Far from taking advantage of the hero-worship he has inspired in the heir to the most powerful throne in Earthsea, Ged instead presents his most prosaic self and dons his most unglamorous disguises, pretending not to notice the disillusionment and even disgust on Arren's part that follow. Most of all, Ged makes no attempt to stop Arren through any mage-power or physical force from following Cob into the realm of the undead/unliving. He merely instructs Arren in the wisdom of self-restraint and the need for courage—most signally, the courage to live a life of insecurity and uncertainty, and to accept the inevitability of death.

Le Guin marks Arren's role as an apprentice in honor by explicitly identifying him with the ancient sword of Serriadh that he carries, the second–most-powerful sword in Earthsea. It is a sword that demands honor and self-restraint from those who wield it: considered "wise" because it is "reluctant to be used,"[18] it never has been drawn, or can be, except to serve life. Arren's given name (as distinct from his true name) means "sword" in the language of his native Enlad, and in fact he was specifically named after the sword of Serriadh. Thus, this sword acts as the objective correlative of Ged's honorable setting-aside of the immense power which he has devoted his life to developing and mastering. At the book's climax, Arren's sword and Ged spend all of their power toward the same anti-immortalist end: defeating the undead mage who has exploited the wall between the living and the dead and is drawing the essence of the world into a void of annihilation.

From entitlement to wisdom

Like Arren, *Tuck Everlasting*'s Winnie Foster is urged in accelerated fashion from what might be called an "honor-neutral" position, where her capacity for self-limitation is yet unarticulated and *in potentia*, to a conscious, ecology-based understanding of honor whose specific function is to act as an anti-immortalist inoculation. Despite her relatively humble position in the book—and in anti-immortalist fantasy—as an ordinary girl living on the outskirts of a small town in the late 1880s, Winnie sits at a unique fulcrum of choice and power. She is the youngest member of the family that has for generations "owned" the woods outside the small town of Treegap, and she is therefore one of the unwitting "owners" of the spring of *elixir vitae*.

Winnie has inherited a tension between a feeling of entitlement toward the small parcel of wilderness that represents the natural world to her and a knowledge of the effect that human rapaciousness has had on it. Her family, having lived in Treegap since the first humans arrived, has borne witness to the heedless deforestation and exploitation of the land. Winnie has heard from her grandmother that the current few acres of trees are all that remain of what was once an endless forest that had surrounded the town, almost entirely cut down by the time period of the book. Winnie's consciousness of this history is heightened throughout the story by the immortal Tuck family's role as multigenerational witnesses. They testify that when they had arrived in the Treegap area eighty-seven years earlier, the small wood "was a forest... that went on and on and on."[19]

At the same time, Winnie has been inculcated with a sense of ownership and a belief that ownership confers impunity. She conveys to Jesse Tuck repeatedly, upon first meeting him, that possession means permission: "It's my wood...[The spring is] mine...if it's in the wood."[20] Her belief that the legal entitlement to drink from the spring is the only factor in deciding whether to drink from it acts as synecdoche for the pioneers' similar exploitation of the surrounding region. Winnie demands to know what other reasons there could be for refraining from using a piece of nature for her own purposes, and much of the next two-thirds of the book is given over to supplying these reasons.

Winnie is more comfortable about these entitled views than the reader is likely to be, for the narrator has already undercut the validity and the wisdom of "possession" in the first chapter of the book, highlighting the thoughtless arrogance of applying such a concept to the living world. "The ownership of land is an odd thing when you come to think of it," the narrator remarks pensively. "If a person owns a piece of land, does he own it all the way down, in ever narrowing dimensions, till it meets all other pieces at the center of the earth? Or does ownership consist only of a thin crust under which the friendly worms have never heard of trespassing?"[21] This Socratic challenge to the sanctity of private property and the concept of human dominion over nature draws attention to Winnie's focus on ownership (or the lack of it) and her reliance on it as a basis for her decisions. Thus, part of her education is to join the narrator and reader in questioning the nature of "possessing" nature; another part is to dismantle the false synonymy between possession and permission.

However, for the most part, her role is observably similar to Arren's in *The Farthest Shore*: to receive an education in eco-spiritual, anti-immortalist honor, so that at the end she can demonstrate the capacity for—and the wisdom of—refraining from following an opportunity for never-ending life. Winnie's section of the plot largely comprises her developing honorable self-restraint toward two applications, one considerably more difficult than the other. The first is that she not tell anybody about the existence and location of the spring of elixir vitae. The second is that she not drink from the spring herself. To these ends, Winnie, like Arren, spends long stretches of several key chapters acting as an audience for lectures on anti-immortalist honor from the two oldest Tuck men, clearly serving as authorial mouthpieces.

The elder Tuck works on the harder of the two rhetorical tasks, convincing Winnie that to become immortal would be an act of deep folly and ineffaceable regret. He describes his unaging, immortal, invincible family as "forever trying to get loose, but stuck";[22] he says that the deep regret of choosing immortality is something a person only feels afterward, when it is too late. He therefore must convince Winnie rhetorically, and in the abstract, before she irreversibly makes the wrong choice.

The elder son, Miles, has the job of persuading Winnie to keep the spring a secret. This is an easier sell than convincing her to accept death as an inevitability when she has the option of living forever. Rather than the eco-spiritual oratory that his father has employed, Miles mounts a case for human overpopulation. After he tells Winnie that frogs only proliferate in the pond when it is relatively free of the snapping turtles that prey on them, she reflects on this peril to the frogs.

> "It'd be nice," she said, "if nothing ever had to die."
>
> "Well, now, I don't know," said Miles. "If you think on it, you come to see there'd be so many creatures, including people, we'd all be squeezed in right up next to each other before long."

Winnie squinted at her fishing line and tried to picture a teeming world. "Mmm," she said. "Yes, I guess you're right."[23]

This simple appeal to an understanding of Malthusian population dynamics suffices to convince Winnie, still thinking about Miles' words some time later, that the Tucks are right in seeking to keep the spring secret—including from the mosquitoes that annoy her on their outing on the pond and would become a plague if they were made immortal. Quite readily, she decides that she will keep the secret of the magical spring.

Mere words are not enough, though, to cement Winnie's decision to restrain herself from drinking from it. She needs, instead, Miles' active demonstration of honor and what could be construed as nobility of purpose that resonates with her own purpose-driven and inherently honorable ambition to make a difference in the world. She finds this desire, which previously she had disclosed only to the befriended toad at home, echoed in Miles' statement that he wants to "do something important." "Winnie nodded. That was what *she* wanted."[24] Her respect for and identification with Miles, who believes deeply in aligning with nature and agrees with his father that immortality is a burden, seems to be what ultimately tips the balance and brings her perceptions and values into line with anti-immortalist thinking.

Later, when she substitutes for Mae Tuck in the prison cell in order to dupe the night guard, Winnie demonstrates her newfound courage and honor all the more in that she never entirely believes the Tucks' story that they are immortal and invincible. All the evidence she receives in this regard is hearsay propagated by the man in the yellow suit and the Tucks themselves; the claims are presented as claims, and remain claims, unconfirmed and partly disbelieved, to the end. Winnie toggles between a credulity based on analysis—for instance, the Tucks' evident relief and spontaneity in telling her the story that she soon suspects they have never told anyone else—and thinking that the tale is nonsense and the Tucks are crazy. Even at the book's climax she is skeptical, wondering if she truly would live forever if she drank the spring water, and, if she did, whether she would regret it eternally, as the Tucks did. She decides, in the end, that it wasn't true, and that the Tucks are probably insane—though she loves them anyway.

Skeptical about the Tucks' immortality, she cannot be sure that Mae Tuck was, as she claims, truly averting world-disaster by killing the man in the yellow suit, and she cannot know if her own actions will save the world as she knows it or only a common murderess. She risks her safety, and forfeits her family's trust, when she could easily discount the gravity of the situation to justify staying safely at home.

The reader, from their position of dramatic irony, knows how much hangs on Winnie's choice, for the omniscient narrator has verified early on that the spring's power to confer immortality is real. In the first chapter the narrator editorializes,

> [If] the people [had] noticed the giant ash tree at the center of the wood, … then, in time, they'd have noticed the little spring bubbling up among its roots in spite of the pebbles piled there to conceal it. And that would have been a disaster so immense that this weary old earth owned or not to its fiery core, would have trembled on its axis like a beetle on a pin.[25]

Some pages later, the narrator remarks that Mae Tuck, her husband, Miles, and Jesse have looked the same for 87 years; similarly, in the Epilogue, she has the Tuck family come back through Treegap, still the same age as in the main part of the book, several years after Winnie has died—so that the reader knows, but Winnie cannot, that she did in fact save the world by helping Mae get out of prison so that the public could not find out about her invincibility.

After abetting Mae Tuck's escape, Winnie demonstrates her now-explicit understanding of honor as a core characteristic when her parents confine her once again to her prison-like yard and monitor her even more closely. Although she has now realized, through having "escaped" twice, that she could slip out every night without their knowledge if she wanted to, she chooses to act honorably, in a conscious way, by deciding not to take advantage of her family's trust in her, and instead to accept their increased monitoring and her tighter confinement without complaint.

More importantly, although the implied storyline between the final chapter and the epilogue tells us that Winnie has lived all her life in Treegap, marrying there and having children of her own, and tells us that the source of the elixir vitae was accessible until the year before her death, Winnie *does* die. In doing so, she demonstrates that she had a self-restraint formidable enough to stave off the continuous temptation of eternal life—not only for herself, but for her loved ones.

The prerequisite of courage

For Arren and Winnie, the joint education in honor and a pro-mortality belief system requires them, first and foremost, to develop courage. This is not the warrior's bravery of a Taran or an Aragorn, but the emotional and spiritual courage without which the fear of death and passion for life will move from normal instincts and outlooks to an obsession. Both books present this heightened level of fear as stemming from human segregation from nature, but it takes different forms: in *Tuck Everlasting* the chief fear is biophobia, while in *The Farthest Shore* it is necrophobia.

Fear of death, of personal extinguishment, drives all of the catastrophic action in *The Farthest Shore*. The inciting incident, the powerful mage Cob's breaching of the gap between life and death, was compelled by Cob's terror at the thought of dying. When the two wizards had first met, Ged tells Arren, Cob "was more afraid of death—of his own death—than any man I ever knew...I have never seen a fear like that; it sickened me with its own sickness."[26] Thus sickened, Ged lost some of his own wisdom, which the trilogy throughout equates with honor and the minimal use of mage-power. Full of anger and hubris, he forced Cob across the stone wall at the edge of death's bourne. This lapse in self-restraint on Ged's part, his momentary willingness to exercise his power advantage, turns out to be fateful, for it makes Cob even more determined to conquer death: "Oh, a lesson you taught me, indeed, but not the one you meant to teach! I have seen death now, and I will not accept it... I will not go that way, I will not cease to be myself! ... I opened the door that had been shut since the beginning of time."[27]

It is fear that gives the immortalist creed power throughout the book. The undead Cob amplifies and exploits the latent fear of death within the human population of Earthsea, persuading artisans and wizards alike to give up their names, their essence, and their power in return for the false promise of a way back from the land of death to that of the living. The former wizard Hare, addled on the opiate drug hazia, gives a rambling explanation of the barter to Ged and Arren: "Remember what the secret is? Remember? No death. No death—no! ...No fear. No death."[28]

Arren, courageous though he is in the manner of princes, is susceptible to this fear, and battles it throughout the book, succumbing on more than one occasion. When he is captured by slavers, his main point of distress is not the mortification of his future lot in life, nor the travails he will undergo, but knowing that he is likely to be dead in a matter of days. Though he understands this and accepts it as a fact, he becomes so gripped with fear that he stops himself from thinking about it. Perhaps because he has been briefly in Cob's presence, or perhaps merely as an inherent weakness of character, Arren's fear of death escalates from this point, until he is immersed in it and at points debilitated by it. He has dreams of being in the land of the dead, on a vast, barren plain full of fear and horror.

Le Guin displays at length the catastrophic outcomes of a fear that has been coaxed into terror and then admixed with horror, for the eventual heir to the high throne of all Earthsea essentially goes mad, for a time, with necrophobia. Unable to rid himself of the nightmares' miasma, Arren is overwhelmed by the fear of death, and his every perception and emotion becomes corrupted. He thinks of his and Ged's quest, which began as one that wakened "the depths of him...not by a game or dream, but by honor, danger, wisdom,"[29] through a distorting haze of doubt and disgust: "Is this how a great peril is met, by sending out an old man and a boy in a boat? … He is mad himself; it is as he said, he seeks death. He seeks death, and wants to take me with him. But I am not mad and not old; I will not die; I will not go with him."[30] As the nightmares continue and his death-phobia seizes him more fully, Arren descends from a courageous and honorable prince to a craven, and he sees the Archmage of Earthsea, the most powerful wizard of his time, as a fool and an imposter. He chastises himself for entrusting his life and heart to a man who is secretive, impulsive, and unwilling to exercise control over his life, even so far as to try and save it.

Through the lens of necrophobia, all Arren's former perceptions are reversed and the wisdom of honor is overthrown. After formerly admiring Ged's honorable self-restriction of his mage powers, Arren now views the wizard as "miserly," and sees his cryptic statements as "grudging, hard to understand. There, thought Arren, lay the very heart of wizardry: to hint at mighty meanings while saying nothing and all, and to make doing nothing at all seem the very crown of wisdom."[31] Having lost his sense of personal honor in his continual terror of death, Arren takes advantage of the power of his princely tone and force of will to coerce speech out of the mad dyer Sopli, who has been driven to yammering incoherence and irrationality by Cob's seductive promise of eternal life. The fruit of this dishonorable exercising of a power advantage is a further contamination of Arren's mind. Sopli affirms Arren's view that Ged is suicidal because he refuses to believe in or be seduced by the promise of eternal life, and will not countenance giving up his name to "fit" through the gap that Cob has made between life and death.

And although at first Arren had argued stridently with Ged against taking Sopli on their quest with them, after this conversation he feels kinship with the madman, going so far in his deranged thoughts as to think that if worse comes to worst, Sopli might fight Ged with him. Arren's character is almost wholly vitiated, his identity corroded, by the time the boat is attacked, Sopli killed, and Ged badly wounded. Instead of responding as a trained prince, Arren feels only a sick horror. Instead of feeling courage rise up in the face of fear, he is full of resentment that any of this is happening.

The turning point of Arren's character arc comes when he overcomes his shame of fear enough to tell Ged of it, and meets the moderating force of the Archmage's wisdom. He explains that he has been so overcome with the terror of dying that he could not look Ged in the face—that he could do nothing but try to hide from death. Ged replies, "There is no safety, and there is no end.… You have strength enough. …Consider, Arren. To refuse death is to refuse life."[32] The acknowledgment of his fear comes more and more easily to Arren for the rest of their quest, and with it courage. Eventually, he can face the heart of the immortalist ruse directly and risk having his cherished hope of eternal life extinguished.

Not noticing that he is now treating Ged as an equal, his hero-worship turned into the respect of equals, Arren challenges the archmage to explain why, knowing that he will die, he should welcome this fact—for himself or anyone else. "You would have me say that [immortality] is not better. But tell me why… . If I love life, shall I not hate the end of it? Why should I not desire immortality?"[33] Arren's new courage lends him the strength to accept Ged's anti-immortalist reply even as he learns, for the first time, of Ged's belief that Arren, as one born to power, might actually be able to achieve Cob's perverted form of immortality. "Life rises out of death, death rises out of life; in being opposite they yearn to each other, they give birth to each other and are forever reborn… In life is death. In death is rebirth. What then is life without death? Life unchanging, everlasting, eternal?—What is it but death—death without rebirth?"[34]

Arren's square facing of the fact of death, his questing for an answer that he does not want, confirms for Ged that Arren has the courage to exercise anti-immortalist honor, and thus to lead the fight against the Unmaking forces of immortalism. Like Arren's old fencing master, whose honorable aim was to make Arren skilled enough to challenge him, Ged shows the incredulous joy and hope of having found, at last, an equal.

Only when Arren has exhausted his fear of fear, when he has been frightened of death too much for too long, become sickened by, and finally rejected it, does death no longer have power over him. And only then, in turn, can he lay claim to his honor, have his identity restored to him, and operate freely and wisely as Agni [King] Lebannen, the heir to the High Kingship of Earthsea.

An antidote to biophobia

Where Ursula Le Guin posits the fear of death as arising from arrogance and power-hunger, and therefore residing at one remove from the human severance from nature, Natalie Babbitt presents a somewhat different—although equally reciprocal and mutually perpetuating—triangulation between fear of death, fear of nature, and lack of honor. Babbitt seems to propose in *Tuck Everlasting* that biophobia, as a potent form of fear, inhibits honor by acting as a preemptive restraining force. The beginning of the novel shows that as long as fear keeps Winnie from doing things even though she can, she will never be in a position to learn not to do things just because she can. In other words, Winnie has had no opportunity to make honorable decisions (or dishonorable ones) because her fear, depicted as mostly being of the natural world, has restrained her in honor's stead.

The Foster family's fearful and antagonistic relationship with nature, and the way it deprives Winnie of the autonomy she needs to develop honor, is highlighted from the beginning. The parameters of her family's home—a house with a "touch-me-not appearance" and an air of complacent pride—are conspicuously marked by a four-foot iron fence. The contrast between the yard inside the fence, with "grass cut painfully to the quick,"[35] and the weeds, dancing gnats, and large toad that reside immediately outside it, emphasizes the fence's role in equating segregation from nature with lack of personal freedom.

Winnie, who makes a habit of sitting on the grass on the inside bounds of her imprisonment and putting her arms through the bars to pull up the weeds on the outside, understands the yard as a cage and herself as a trammelled creature. After suggesting to the toad that she might make a pet of it and put it in a cage with lots of grass, she then relents, remarking that then the toad would be as much a prisoner as she is. "Why should you have to be cooped up in a cage, too? It'd be better if I could be like you, out in the open and making up my own mind. Do you know they've hardly ever let me out of this yard all by myself?"[36] After she has been "kidnapped" by the Tucks and then returned home, and has had experiences that belong only to her, separate and secret from her family, she is consciously aware of the yard as a prison as she grips the bars of the fence and thinks about Mae Tuck being "behind another set of bars in the jailhouse."[37]

Adding to her physical confinement is a relentless surveillance by her guardians, which frequently involves cautions against any enjoyment of the natural world. Her parents and grandmother are oddly disembodied beings throughout most of the book, often appearing as voices warning Winnie of the distasteful consequences of being in any contact with nature: "Winifred! Don't sit on that dirty grass. You'll stain your boots and stockings." "Come in now, Winnie. Right away. You'll get heat stroke out there on a day like this."[38] "A toad?...Nasty things, toads."[39] Babbitt explicates this motif when Winnie enters the Tucks' chaotic and untidy home and is shocked by its disorder. Wryly, but not unsympathetically, Babbitt describes how Winnie had grown up with the "pitiless" order of her mother and grandmother's housekeeping, which had left the house itself in a state of "limp submission." And, she notes, "Winnie was in training"[40] to be this same kind of nature-denying, dirt-fearing adult.

The result is that although it is partly Winnie's obedience that keeps her in a yard that she can actually walk out of at any time, fear plays a larger role.

> [T]he world was a dangerous place. People were always telling her so. And she would not be able to manage without protection. They were always telling her that, too. No one ever said precisely what it was that she would not be able to manage. But she did not need to ask. Her own imagination supplied the horrors.[41]

When Winnie does slip out of the yard the next morning and go into the nearby woods which the Foster family has "owned" for generations, Babbitt shows how biophobia and necrophobia stem from the same need to avoid or eliminate fear rather than meet it. Winnie experiences a disproportionate and naïve relief when, in the immediate term, the world outside the fence appears far less fearsome than she had thought. The wood, to her surprise, is beautiful, and she wanders through it for a good while, taking it all in, "proud to forget the tight, pruned world outside."[42] Smelling the lovely, damp air and noting the quality of the green and amber light coming down between the trees, she soon wonders why she had never gone to the woods before. Once she is kidnapped by the Tucks, and taken involuntarily into the wider natural world, "The sweet earth open[s] out its wide four corners to her like the petals of a flower ready to be picked, and it shimmer[s] with light and possibility until she [is] dizzy with it."[43]

As Winnie moves from the entombment of her house to the full dimensionality of the world, her new passions are almost immediately perverted by proximity to immortalism. Upon hearing the Tucks' story, her thoughts move to the possibility that she might cling to her rapturous newfound life and eliminate any lingering fears of the wider world by becoming both immortal and invincible, like the Tuck family. The idea of living forever in her newly discovered world fills her with joy, and she races down the road shouting, elated and for once uninhibited.

However, Tuck soon desperately presses his case for the disastrousness of immortal life—for her individually, for humankind, and especially for the earth. Winnie is paralyzed at this forced confrontation of the ultimate fear.

> She—yes, even she—would go out of the world willy-nilly someday. Just go out, like the flame of a candle, and no use protesting. It was a certainty. She would try very hard not to think of it, but sometimes, as now, it would be forced upon her. She raged against it, helpless and insulted, and blurted at last, "I don't want to die."[44]

Immobilized, like Arren, by her anguished struggle with the inevitability of death, Winnie can only sit, huddled up and depressed, her bubble of elation burst.

As with Arren, Winnie's natural, latent fear of dying is artificially amplified by the presence of the "hope" of immortality and the campaigns of the story's two immortalists, Jesse Tuck and the man in the yellow suit. These two figures eventually distort her perceptions and poison her feelings toward the Tucks, in turn barring her access to their anti-immortalist wisdom. In place of the sense of liberation she had earlier felt at her kidnapping, she now feels it to be an outrage that she has been taken by the Tucks and forced to stay in their messy hovel. She sees the Tucks as criminals, and the man in the yellow suit—who had formerly reminded her uneasily of the black ribbons they had hung on the cottage door after her grandfather died—as her only savior.

Only when she sees death firsthand, witnessing Mae's death-blow to the man with the yellow suit and surviving her own horrified fear of it, does Winnie begin to find courage and thereby regain both her perspective and her honor. She aligns herself with the Tucks, telling the Treegap constable that they had not kidnapped her but that she had come of her own free will. She then embarks on the scheme to rescue Mae Tuck from incarceration and hanging—a hanging that Mae will survive—to keep hidden "[t]he horror that proved the secret" (119).

Confident in her ability to face and withstand the ultimate fear, Winnie rids herself of immortalist creed by emptying the vial of *elixir vitae* over her friend the toad, who seems permanently imperiled by wagon wheels, drought, and roaming dogs. Although she ostensibly staves off her decision by reminding herself that she can always go to the spring and get more, the reader is hardly surprised when the Tucks, returning to Treegap a few years after the Second World War, find her epitaph in the cemetery: "Winifred Foster Jackson / Dear Wife / Dear Mother / 1870–1948."[45]

The immortalist as horror figure

It is no coincidence that both *The Farthest Shore* and *Tuck Everlasting* belong to the rhetorical mode of fantasy that theorist Farah Mendlesohn terms "intrusion fantasy," in which "the world is ruptured by [an] intrusion, which disrupts normality and has to be negotiated with or defeated, sent back whence it came, or controlled."[46] In order to make clear the full disruptive nature of the intruding element, Mendlesohn explains, "the base level is the normal world, [and] intrusion fantasies maintain stylistic realism and rely heavily on explanation."[47] Significantly for the understanding of these two books, intrusion is the key modality of horror fantasies, just as virtually all fictions that portray human immortals—vampires, zombies, ghouls—also belong to the realm of horror.

It takes little adjustment of perspective to see *The Farthest Shore* and *Tuck Everlasting* as intrusion horror stories, although this tone pervades to different degrees in each book. The Earthsea trilogy also belongs to another of Mendlesohn's four fantasy categories, that of the "immersive fantasy," in which the fantastic is presented "without comment as the norm for both the protagonist and reader" and, like science fiction, uses "the irony of mimesis," with the secondary world "acquir[ing] a scientific cohesion all its own."[48] Yet, as Mendlesohn points out, this very establishment of a fully created, mimetic, "normal" world makes it possible, even common, for immersive fantasies to hold within them intrusion fantasies that function exactly as non-immersive fantasies do, driven by events that transgress the rules of the secondary created world. This is exactly what happens in *A Wizard of Earthsea* and *The Farthest Shore*, with human immortality acting as the intruding (and horrific) element.

Both the chief antagonist of *The Farthest Shore*, the untrained but powerful wizard formerly known as Cob (but now nameless and thus self-less, as Ged points out in a long climactic confrontation), and the villain of *Tuck Everlasting*, the similarly nameless "man in the yellow suit," are in essence anti-villains: craven, sly, unmanned figures whose ability to dominate lies almost solely in their dishonor, their readiness to take advantage of others' weaknesses, among which both men count their other people's honorable self-restraint. Unlike Lloyd Alexander's Arawn, C. S. Lewis's Tash and White Witch, Tolkien's Sauron, or other fantasy antagonists, Cob and the man in the yellow suit do not confront powerful protagonists, nor conquer through effort, strength, or superior qualities. Instead, they achieve their positions by capitalizing on what now would be called sociopathic characteristics.

Babbitt describes the man in the yellow suit as a combination of salesman and puppet-ghoul, with the former's unctuous comportment and constitutional willingness to breach social etiquette and the latter's soullessness. Babbitt extends the alliance from atrocity to monstrosity by portraying him as an "inhabited" body with only awkward, mechanistic control over his physical being. He is a malign puppet: his tall body moves constantly, with tapping feet, twitching shoulders, and spastic jerks—yet also "a kind of grace, like a well-handled marionette."[49] He is unnaturally white, a corpse-like figure, with the long, thin fingers of a prestidigitator. While negotiating with the Fosters for Winnie's return, his foot begins jiggling of its own accord, and when he confronts the Tucks and Winnie the next day, telling them of his intentions to sell the *elixir vitae*, his face at first is smooth and empty, but then begins to loosen, in an image of a sloughing mask; "his eyes [glow]...and a nervous petulance [comes] into his voice."[50] When Mae Tuck knocks him out, he is described less as having lost consciousness than having been severed from a manipulating source: He lies on the grass, motionless, and looks "more than ever like a marionette, a marionette flung carelessly into a corner, arms and legs every which way midst tangled strings."[51]

The man in the yellow suit gains the strategic advantage by hiding his purposes and extracting information and concessions from children, the distraught, and the ignorant. On his first appearance, emerging from the gloaming with a slight glow emanating from his yellow suit, he presses 10-year-old Winnie for information when she is alone in the yard and twice keeps her from calling her father out of the house. When her grandmother, suspicious at the sound of conversation, does come out and tries to take charge of the situation, asking the stranger who he is and what he wants, he answers neither of her questions, thus retaining his power position.

The next time he sees Winnie's family, he is operating from the advantage of knowing where their kidnapped daughter is, knowledge that he ruthlessly trades for the deed to the woods that hold the spring of immortality. Dishonestly framing the exchange as an equal he tells the family that he has what they want and they have what he does. The transaction completed, he puts the cap on his dishonorable behavior by recasting himself as the savior, telling the Foster family with satisfaction, "There now! Your terrible ordeal is as good as over, isn't it? I'm so thankful I was here to help you out!"[52] Babbitt heightens his domination by writing the whole exchange as a monologue by the man in the yellow suit, leaving the Fosters' side of it off the page and thereby effectively muzzling them.

In the presence of authority figures, the man in the yellow suit resorts to the simple expedient of lying. He counters the constable's suspiciousness by claiming that the Fosters are friends of his who have sold him the wood; then, as the two ride toward where the Tucks have taken Winnie, he strips the constable of power by remaining almost mute in the face of the lawman's gregarious disclosures.

His downfall is in mistaking the Tucks' honor for weakness, evaluating their behavior by his own perverted morality. After he has offered the Tucks an opportunity to work for him as an "exhibit" proving that the spring's waters confer invincibility, only to have them denounce him as "a madman"[53] he sneers at their "selfishness" and "stupidity" in keeping the spring secret instead of exploiting it as he intends to. "Now it's too late," he tells them. "There's nothing you can do to stop me."[54] Believing that their self-restraint stems from lack of initiative, he disregards the fact that they outnumber him and have a gun. Mae Tuck, the Tuck family's sensible-minded matriarch, hits the man in the yellow suit with the butt of her rifle, knocking him out with a blow from which he never recovers.

Intimations of a wider-scale dishonor come with Babbitt's none-too-subtle links between the man in the yellow suit's elitist thinking and eugenicist beliefs. His use of terminology like "barbarian" and references to the Tucks as "rough country people...illiterates"[55] achieves explication when he tells the Tucks and Winnie that he would distribute the *elixir vitae* selectively: not to "just anybody," but only to the deserving, educated rich. "Ignorant people like you should never have the opportunity. It should be kept for...certain others. And for me" (ellipses in the original).[56]

In this, of course, Babbitt alludes to a long-running quarrel between fantasy and science fiction about whether power discrepancies can be avoided in technologies and discoveries that allow unnatural longevity, whether there is an inherent potential for great dishonor within human society were such technologies to become concentrated in the hands of the powerful, and about the inequities that already exist in access to life-preserving medical care. But she also brings in the shade of Mengele, and allies the immortalist's readiness to exploit with Nazism and the Jewish Holocaust.

The other immortalist in *Tuck Everlasting* presents a subtler and perhaps more disturbing kind of horror. Jesse Tuck, seen through Winnie's eyes as a "glorious," "wonderful," "beautiful" boy,[57] is the only member of the Tuck family who enjoys his immortality. In discounting his father's sermons about "the wheel" and the circle of life, and disdaining his family as prudish for their deeper and grimmer perspective of immortality, Jesse actively sabotages his family's anti-immortalist campaign by painting a much sunnier picture of eternal life and physical impunity. His glowing depiction of immortality and invincibility provokes his brother Miles to stricture, "That kind of talk'll make her want to rush back and drink a gallon of the stuff...There's a whole lot more to it than Jesse Tuck's good times, you know."[58]

Jesse at first seems a sympathetic character, and, indeed, he precipitates Winnie's kidnapping by trying earnestly to keep her from drinking the water of eternal life. However, he soon strikes an unsettling chord by sexualizing Winnie despite her prepubescent status. After she asks him if he's married, he asks her in return. "'Of course I'm not,' she said. 'I'm only ten. But I'll be eleven pretty soon.' 'And *then* you'll get married,' he suggested."[59] Although the 17-year-old Jesse's quickness in capitalizing on and encouraging Winnie's infatuation may initially seem like a holdover of the sexually naïve zeitgeist peculiar to the 1960s, further events suggest that Babbitt is using Jesse's enjoyment of the inappropriate sexual dynamic to signify the corruption of the immortalist character.

On several occasions, Jesse pushes Winnie toward drinking from the spring with a promise that if she does it when she is 17, they can get married and she can live forever as his wife, gallivanting around the world and seeing all the wonders that Winnie has just become aware of. "[Y]ou and me, we could have a good time that never, never stopped."[60] Winnie's deep loneliness, desire for love, and infatuation with Jesse make these powerful enticements. In fact, even in the novel's final chapter, Winnie is still thinking, "[I]f all of [the Tucks' story about the spring] was true, there was more water in the wood. There was plenty more. Just in case. When she was seventeen."[61]

In his refusal to see "more to it than Jesse Tuck's good times," his resistance to experiential wisdom, and the vanity and frivolity that have him constantly showing off for Winnie, Jesse demonstrates the destructiveness of immaturity, as well as underscoring the novel's nature as a meta–coming-of-age story—for of course neither Jesse nor Winnie can come of age unless they age to begin with. He also represents another face of immortalism, that of the careless sociopath whose motivations are not glory or profit, but sheer selfishness. Late at night, dishonestly, he sneaks up to her bedroom to seduce Winnie into joining him in immortality, coaxing her to drink the water when she is the same age as him and becoming his partner in self-indulgent eternal life.

It is worth noting that, ultimately, Jesse poses a considerably greater threat to the state of the world than does the man in the yellow suit, for Winnie is conspicuously hungry for love, warmth, and companionship, and Jesse is more than willing to exploit this susceptibility. Through Jesse, Natalie Babbitt seems to be alerting the reader to the more glamorous and innocuous-seeming vehicles for the disastrous creed of immortalism.

In her antagonist, Cob, Le Guin reifies immortalist thinking as physical monstrosity, connecting the character trait of dishonor to tropes of horror that at times nearly reach caricature. Cob exhibits his sacrilegious nature when he makes his living using the Lore of Paln, which allows magic-workers to manipulate the borders of life and death, to summon the spirits of great Earthsea heroes from the dead as a kind of party trick. Ultimately Cob finds a way to cross freely between life and death, and to bid the souls even of great mages and proud women to toggle back and forth at his prideful command. He gains control over the dead spirit of Ged's predecessor, the Archmage Nemmerle, making "that great soul come at his call, like a dog to heel,"[62] and even over that of Earthsea's greatest king and warrior, Erreth-Akbe.

The great mages of Earthsea have known of Pelnish Lore for a thousand years, and in that span, a lifelong indoctrination in the voluntary restraint of their power has kept those with sufficient mage-craft from using it. This honor has kept Earthsea from ruin, Ged tells Arren, for even a king would not be able to restrain a great mage from using his power if he chose otherwise. He points out that even in the recent centuries where Earthsea has had no high king, the Archmages "kept fealty and served that common law" of self-restraint.[63] To use the Lore of Paln and manipulate the spirits of dead people is within the mages' power, says Ged to Arren, "But it seldom done, and I doubt that it is ever done wisely."[64]

Thus this deep peril to "the great Balance" stays checked only by a tacit agreement that it is evil and open to exploitation by any wizard with both the inherent capability and a lack of honor. The wizard Cob has both of these. Reckoned by his community to be only a sorcerer, he in fact has, as Ged explains, the inborn power of a great mage. True to the axiom that to keep the mageborn ignorant and untrained in honor is a dangerous thing, Cob remains uneducated in the reasons for self-limitation and allows to grow unchecked within him what Ged calls "the self that cries *I want to live; let the world burn so long as I can live!*"[65]

Like Babbitt's man in the yellow suit, Cob is narcissistically unable to assign to others more honorable motives than he himself possesses. He boasts to Ged and Arren, when they find him, that he, among all mages, has found the way to immortality. When Ged replies, "Maybe we did not seek it,"[66] Cob does not believe him, going on to accuse Ged and his mages of talking about acceptance and the balance of life and death to hide their failure to discover the secret of immortality.

That courage, not immortality, is the only non-catastrophic response to the natural fear of death is made clear by the supposedly immortal Cob's continuing state of terror. "I have no fear," he claims when Ged and Arren find and confront him on the edge of the world. "What should a dead man fear?" Ged replies, "Surely not death? Yet it seems you fear it."[67] Even as the self-anointed "Lord of the Two Lands,"[68] Cob is reduced to a gibbering, cowering figure by his terror of the idea of his individual consciousness ending.

In failing to receive, or perhaps failing to accept, what E. O. Wilson would call the education that allows a person to put "a high premium on posterity,"[69] Cob reaches an extreme of humanist arrogation and denies his place within the natural world and the Great Equilibrium. One might suspect Le Guin of modeling Cob on Alan Harrington, so much do his monologues sound like passages from that author's book *The Immortalist*: "Let all stupid nature go its stupid course," says Cob, "but I am a man, and *better than nature, above nature*. I will not go that way, I will not cease to be myself!"[70] (emphasis mine). In rejecting self-limitation, Cob begins the Unmaking of the world, ending Ged's life as a mage and nearly bringing annihilation to the whole of Earthsea.

Le Guin presents the immortalist as a whited sepulchre, with a glamorous, seductive face overlying a core of ghoulishness. Cob's "seeming," the image that he sends back from the land of the dead, is a product of vanity, appearing to Arren as an imperious, charismatic, Christ-like figure, tall and enshadowed, holding a flame in his palm like a pearl to offer Arren never-ending life. When Ged and Arren finally reach the shore on the edge of the world where Cob's undead body resides, he sends a more glamorous image yet, as a "strong, comely" man with long black hair and a visionary mien.[71]

However, when the great dragon Orm Embar expends his life in "killing" Cob's sending to save Ged from death, the revenant appears as he truly is, as a figure whose attractiveness is illusory and whose true self has been turned monstrous by his immortalism. When he is revived from the dragon-caused "death," he is ugly and shriveled, crawling spiderlike over the sand dunes. The physical death that had been visited on him years earlier shows in the empty eye sockets, the withered mouth, and the desiccated skin. By the end, even Cob's immense mage-power cannot keep his perversion of the laws of nature from translating to a perversion of what remains of his corporeal being. The hollowness of the promise of eternal life manifests in his increasing inability to restore his physical image; when Ged forces him to accept that immortality is not the opposite of death, but the opposite of life, "the blind man … looked as if he wept, but he had no tears, having no eyes. His mouth opened and shut, full of darkness, but no words came out of it, only a groaning."[72]

The balance and the cycle: ecological anti-immortalism

The paradigms within which Le Guin and Babbitt pit honor against immortalism bear the hallmarks of Arne Naess's concept of Deep Ecology and touch the edge of E. O. Wilson's biophilia hypothesis. They extend scientific ecology's notions of biotic and abiotic interdependence to infer an ineluctable human connection with the natural world; they depict their human characters as finding a sense of peace and fulfillment when they are perceiving and living within this spiritual-evolutionary worldview, whether consciously or not. The books' objection to human bodily immortality lies not in what it does to the individual soul or even the nature of human society, but in how such a transgression of nature would rupture the fabric of creation and introduce nihilism to the world. Although both books present humans as arrogating themselves above the rest of the living world (or, in the case of *The Farthest Shore*, the entire created world, including the non-living one), they argue that the human identity as part of the evolutionary and ecological scheme means that their efforts to escape it threaten to destroy the entire structure.

Tuck Everlasting, whose target audience is younger and scope more limited, figures life as a cycle and time as a wheel, the one within the other, so that each element rises and falls in a perpetual pattern of renewal and decay. In what may be the novel's most overt statement of its pre- or un-Christian belief system, the authoritative Angus Tuck theorizes that immortality, rather than being a prelapsarian reward for obedience to God, is "something left over from—well, from some other plan for the way the world should be…Some plan that didn't work out too good."[73] In the revised and better dispensation,

> Life [is] moving, growing, and changing, never the same two minutes together … Everything's a wheel, turning and turning, never stopping. The frogs is part of it, and the bugs and the fish and the wood thrush, too. And people. But never the same ones. Always coming in new, always growing and changing, and always moving on. That's the way it's supposed to be. That's the way it is.[74]

In order to be part of the living world rhapsodically described between segments of

Tuck's homily—the water gliding and slipping and foaming, fringed by clumps of willows, orchestrated by the voices of frogs and caroling wood thrushes—humans must be willing to depart it graciously like its other participants. Dying is part of the wheel, Tuck tells Winnie, just like being born. "You can't pick out the pieces you like and leave the rest. Being part of the whole thing, that's the blessing…Living's heavy work, but off to one side, the way we are, it's useless, too. It don't make sense. … You can't have living without dying."[75]

Many ecocritics have marked the cyclical concept of time and life as the key attitudinal difference toward the living world made by the Abrahamic religions during the Enlightenment. Lynn White, Jr., in "Historic Roots of Our Ecological Crisis," describes Judeo-Christianity's "concept of time as nonrepetitive and linear"—which he holds in "sharp contrast" to the pre-Abrahamic religions' "cyclical notion of time"—as a central part of the "greatest psychic revolution in the history of our culture," one that gave Western societies the psychospiritual permission to exploit and degrade the natural environment.[76] Babbitt restores the pre-Christian, a-Christian, arguably anti-Christian understanding of the natural world from the very first sentence of *Tuck Everlasting*: "The first week of August hangs at the very top of summer, the top of the live-long year, like the highest seat of a Ferris wheel when it pauses in its turning."[77]

The cycle motif continues to the end, with a 25th chapter seeming to be placed as an "unnatural" additional hour to the novel's clock, and the wheels of the Tucks' wagon, like the tires of the highway's cars, just passing by the toad upon which Winnie poured her water of eternal life. Within that enclosing wheel, Babbitt depicts life as composed of illimitable interactions that need no purpose beyond their being, all of them forming a complex interweaving from which no creature could extricate itself without rending the whole. Angus and Miles Tuck's homilies, focusing on the "wheel," are augmented by the narrator's broader, deeper portrait of a multisymbiotic, organismic earth capable of "trembl[ing] on its axis like a beetle on a pin"[78] if the natural law of death is transgressed.

Humans, in *Tuck Everlasting*, consider themselves to be above this busy, intrinsically meaningful matrix. But the novel portrays each step in their removal from it as marked by paucity and impoverishment. Even in the 1880s, at the time of the book's main action, the areas that are held as people's property are notable for their sense of uneasiness: the sun in these places is too hot, the dust oppressive, and the grass itself dismal. At the end, the Tucks return to a town shockingly and tragically changed. The once-otherworldly woods, formerly "a dimness shot through with bars of sunlight, a great many squirrels and birds, a deep, damp mattress of leaves on the ground,"[79] have been bulldozed and the ancient cow path has become a blacktopped road. The town of Treegap itself is now a hodgepodge of utilitarian businesses such as a dry-cleaner, hot-dog stand, gas station, and a chrome-gilded diner. The Tucks continue to move changelessly through a human world increasingly diminished by its inhabitants' foolish striving to "pick out the pieces [they] like and leave the rest."

Where Babbitt's wheel conception of time and life implies an element of balance, Le Guin's novel makes balance the chief component. The Great Equilibrium of the Earthsea trilogy enlarges the theory of ecosystem stability (a theory which, the word "stability" notwithstanding, acknowledges the continual ebbing, fluxes, booms, and extinctions) to the whole of the organic world, living and non-living alike. For considerable stretches of the trilogy's first and third books, Le Guin urges upon the reader a deep-ecological treatise in which meaning (though not necessarily purpose) resides in all elements of creation. For instance, when the young Ged asks his mentor, the mage Ogion, what the uses are of the fourfoil plant, Ogion replies, "You must know it in all its seasons root and leaf and flower, by sight and scent and seed[;] then you may know its true name, knowing its being: which is more than its use. What, after all, is the use of you? or of myself? Is Gont Mountain useful, or the Open Sea?"[80]

The Farthest Shore condenses and reiterates both *A Wizard of Earthsea*'s essentialism and the more specifically ecological belief in nature's ability to self-regulate and self-right. The normally anti-demagogic Le Guin, who has objected to C. S. Lewis's Narnian series on the basis that "as soon as wishful thinking or a conscious political or didactic purpose intrude on [a story's] credence, they deform it and the story loses plausibility,"[81] allows Ged in *The Farthest Shore* expository speeches of remarkable length and frequency on the subject of the Balance and the threat posed to it by human immortalism. "The winds and seas, the powers of water and earth and light, all that these do, and all that the beasts and green things do, is well done, and rightly done," he tells Arren in one such monologue. "All these act within the Equilibrium. From the hurricane and the great whale's sounding to the fall of a dry leaf and the gnat's flight, all they do is done within the balance of the whole."[82] He refutes Arren's anthropocentric view of illnesses and blights as undesirable, explaining them instead as an ecologist would, in terms of their niche role: a pestilence is part of the great equilibrium, which may make earth suffer as it helps to restore balance but is not a reason to lose hope or reject the fundamentals of nature.

In the *Farthest Shore*'s iteration of the Earthsea cycle's essentialism, Ged proposes that humans alone can step out of nature—and even they can do so only if fear and dishonor, harnessed by great inherent power, widen the gap between them and the other elements of the world. He articulates this separation as being not, as in the Judeo-Christian teleology, a vertical one of innate superiority, but a horizontal one born of consciousness, and more particularly the foreknowledge of death. While this foreknowledge imparts to humans the great gift of selfhood, it is also a continual liability, making humans alone capable of evil and making it necessary for them to learn to be natural, where all other elements of nature know this without learning. "Having intelligence, we must not act in ignorance. Having choice, we must not act without responsibility."[83]

In keeping with the book's persistent theme of ecology-based honor, Ged argues that the action most often required for humans to do their part in keeping the balance is to exercise self-restraint, and not act at all. When Arren probes him about the worth of doing nothing, and argues that people must act even if they don't know what the consequences will be, Ged replies, to the prince whom he hopes to soon anoint the High King of Earthsea,

> If there were a king over us all again, and he sought the counsel of a mage…and I were that mage, I would say to him: My lord, do nothing because it is righteous or praiseworthy or noble to do so; do nothing because it seems good to do so; do only that which you must do and which you cannot do in any other way.[84]

It is the gap between humans and other existing things, the gap created by consciousness of death, that admits immortalist fear and threatens to become the void of the Unmaking. *The Farthest Shore* urgently shorthands the essentialist conception that connects being to identity, identity to a "true" name, and name to life. The immortalist mage Cob unravels this thread of being by promising eternal life in exchange for name, though in fact those whose selves Cob has taken are trapped in a Virgilian purgatory until Ged releases them.

A deep awryness

The degree to which Babbitt and Le Guin ground their arguments against immortalist thinking in a belief in biophilia and deep ecology, and thus offer the child reader alternatives to the Christian-based interpretation of human exceptionalism, shows in the way the novels treat not in right and wrong, but in rightness and wrongness. The protagonists in *Tuck Everlasting* and *The Farthest Shore* experience a sense of alignment when they are in what Christians would call a "right relationship" with nature. Conversely, they experience a deep sense of disturbance when they are practicing or confronted with affronts to their evolutionary instincts—and especially the ultimate affront, the unnatural tenets of immortalism.

For Winnie, this sense of "rightness," of the harmonious and life-giving workability of the ecological dispensation, initially sits in conflict with her more orthodox, human-society–based understanding of morality. As someone who has lived with the cleanliness-worshipping, nature-fearing personalities of her mother and grandmother, Winnie has been indoctrinated to feel her family's obsessively clean, anti-biotic way of life as "right," and is "in training" to reject any form of disorder, including and especially the perceived disorder of nature.

The struggle between this indoctrination and her affinity for nature emerges early on in her relationship with the toad across the road, whom her grandmother regards as "nasty" and for whom Winnie feels an obligatory revulsion and contempt, but whom she nonetheless not only befriends but admires for its autonomy and courage. It is, in fact, her anxiety to retain the toad's real or imagined esteem for her, symbolizing her inner sense of the "wrongness" of a life segregated from nature, that tips the balance between the biophobic and the biophilic and pushes Winnie to "run away" to the woods, thus initiating her re-education into alignment with nature.

Winnie's new perceptions and sensations are enforced when she subconsciously absorbs the balance and fecundity of the natural world in and around the pond outside the Tucks' house: the bullfrog and small fish, jewel-like dragonflies, and gleaming, rainbow-scaled trout; the silvery songs of the wood thrush and swarms of insects; the black, silky water against the sides of the rowboat. She relapses later that evening, when, alienated by the Tucks' "revolutionary" indifference to the concept of tidying up[85] and by the absence of her own nightgown and usual bedtime routine, she again becomes prey to her habitual biophobia. The joy she had felt on the open road that morning disappears, and her former fears of the wide world roll back in. But Babbitt shows nature to provide its own antidote to the fear of nature and of uncertainty: as the moon rises, Winnie listens to the comfortable conversation of the frogs, the song of the crickets, and the rustling of a mouse in the drawer of the table, as he eats flapjacks that Mae Tuck has put there for him. Eventually, these things are "clearer in Winnie's ears than the voice of her thoughts. She began to relax, listening to the sound-filled silence."[86]

In these lushly woven passages, Babbitt shows Winnie intuitively learning the "wrongness" not only of the rigid, fearful orderliness of her home and family, but of the man in the yellow suit's fundamentally anti-ecological assertion that "all magnificent things [are] very simple."[87] By the time the man in the yellow suit is "kidnapping" her back and taking her over the same terrain she had covered when a "captive" of the Tucks, Winnie has learned the responsibility she bears in return for the beauty and potential of the earth: When she once again sees the wide world in front of her, she thinks not of her own possibilities within it, but of the possibilities of what might happen to it if a dishonorable immortalist gains power over the water of eternal life.

By now, Winnie has lost the tendency toward self-deception that had marked her earlier. She has, the narrator informs us, a strong sense of rightness, so while she knows that she *could* tell her family that they had never forbidden her to help Mae Tuck escape from prison, it is a silly and specious argument, as of course "it would never occur to them to include such a thing on their list of don'ts."[88] She has also, as importantly, lost her sense of entitlement. In her new perceptions, the wood is no longer *her* wood, but *the* wood. The belonging she experiences is reversed: instead of the earth belonging to her, she has come to understand that she belongs to it. The natural world, personified from the beginning—for instance, in the cows engaged in "slow chewing and thoughtful contemplation of the infinite,"[89]—is now something Winnie is highly sensitized to. When she can't find the toad to pour water on during the height of the August drought, she tips the water instead into the earth, whereupon it is instantly sucked up in a proto-Gaiaistic image of a sentient and thirsty planet. The sun's power, at the end, "a roar without a sound,"[90] has the ability to defeat them all: the woods, countryside, and town, and even Winnie's nature-conquering mother and grandmother, who sit helpless and plaintive in the parlor with their hair in disarray and bodies lax with heat.

Ged and Arren, at first knowing only that some form of evil is leading wizards to forget their powers and harvests to be blighted, literally follow a trail of increasingly strong "wrongness" across all of Earthsea to find the source of the wound, the mage Cob's path out of the land of the dead. On the Isle of Roke, insulated from the world by mage-power, the intimations of evil come to them as faint rumors and an uneasy sense that power and resolution are waning, and the sun itself dimming. In the seas near Roke, both men are capable of praising and rejoicing in the sunrise of the spring equinox.

But their first destination, Hort Town, where Cob has recently worked his bargain of endless life for identity, has an atmosphere thick with awryness. Much of the population sits motionless, numbed by the opiate drug hazia; the magic-workers have lost their belief in magic; causeless fights break out in the bazaar; the artisans have lost not only the knowledge of their craft but the passion for it. Arren is infected by the wrongness, feeling his pride and anger amplified, unable to shake a sense of petulance, nervousness, and disgust. When, after losing his footing in uncharacteristic clumsiness, he says in a rage, "Oh, curse this town!," Ged replies dryly, "No need to, I think."[91]

The trail of disorientation and senselessness grows progressively stronger and more tangible as the wound of the world, Cob's vacuum-like breach in the wall of stone between the bournes of the living and the dead, turns all things nonsensical. Ged and Arren follow Cob's trail of malignity down what Ged describes as "a track of wreckage, as if we followed a runaway cart down a mountainside and watched it set off an avalanche."[92] In Lorbanery, a former hub of artisanal work, the inhabitants have lost their craft and even their knowledge of long-established ballads that create continuity between the eras of Earthsea.

When Ged and Arren reach the floating city of the raft-folk, their chief tells Ged that their position on the open sea isolates them from the wrongness, madness, and unmaking—yet even there, the ageless winter solstice ritual of singing in the lengthening of days comes to a halt when the singer forgets the song, in a portent of the Unmaking. "There are no more songs. It is ended," the singer tells the amazed chief.[93] Arren sees a dimming of the stars and feels that the sun will not rise that day. Dis-ease turns to disintegration: on the Isle of Roke, the Master Chanter likewise stops in the middle of a song, telling his students, that he no longer knows what it means, and the Master Healer finds that his herbs have lost savor, his spells their virtue. The Master Summoner, looking into the island's immanent Fountains of Shelieth, sees "the streams run dry, and the lips of the springs of water draw back....I saw the Unmaking."[94]

As they close in on Cob, near the "farthest shore" of Earthsea, the transgressions become atrocities. The great dragon Orm Embar must seek Ged's help when his kin begin to go mad, losing their capacity for speech and even beginning to eat each other, in a gross violation of their innate nature. As Orm Embar explains to Ged, and he in turn relates to Arren, they have been struck dumb, in both senses of the word, by fear wrought by the undead Cob.

> This man has been among them, having no fear of them, for though killed he returns from death in his body, alive. Therefore they fear him as a creature outside of nature. Their fear gives his wizardry hold over them, and he takes the Speech of the Making from them, leaving them prey to their own wild nature. So they devour one another or take their own lives, plunging into the sea—a loathly death for the fire-serpent, the beast of wind and fire.[95]

Orm Embar tells Ged of other acts of senselessness and wantonness scattered throughout Earthsea: villagers on one island killing a baby on an altar, villagers on another stoning a sorcerer to death. Ultimately, as the dragon guides Ged and Arren through the land of the dead to where Cob is, he too loses his power of speech. He is left dumb, stripped of his knowledge of the Making and his deep and ancient dragon wisdom.

Assumptions of unorthodoxy

It is fitting, and probably necessary, that Winnie Foster is dead at the end of *Tuck Everlasting*. And it is equally fitting that Ged has wholly spent his wizardly powers, thereby relinquishing his individual identity as both he and we have known it throughout the Earthsea trilogy, at the end of *The Farthest Shore*. Le Guin explores this identity loss vis-à-vis masculine egoism in her continuation of the series in the early 2000s, but it seems clear that initially the identity-extinguishment was meant as a figurative death, especially given that Le Guin bases Ged's strong charisma almost solely on the extraordinary strength of his powers and his status as one of the greatest wizards in Earthsea history.

At the same time, the fact that the figurative or literal deaths of two chief protagonists, rare in modern-era children's literature, have passed unremarked by critics or gatekeepers suggests the degree to which adult society seems willing to let implosive messages go by if they abet the expected affirmation of the child's allegiance to nature. Winnie's death has failed to stir the controversy of, say, Leslie Burke's drowning in Katherine Paterson's *Bridge to Terabithia*. And although, granted, the event happens rather more "off-stage" than Leslie's, occurring between the final chapter and the epilogue, surely her death's thesis-driven nature as an implied rebuke of immortalism and technological "progress" makes it more consequential than the theme-driven death in Paterson's book.

Similarly, Ged's voluntary extinguishment of his wizardly power has to my knowledge occasioned no discussion about the book's appropriateness for the reader "age 11 and up" designated by the publisher. Yet it seems likely that his action's unsubtle indictment of the increasingly vigorous campaign for extended life, youthfulness, and individual achievement affects the young reader as profoundly—if differently and less consciously—as if he had died in battle.

The two novels are, indeed, notable in showing how much children's literature can get away with in imparting radical ecological values. Their case against human bodily immortality confrontationally replaces a set of standard Abrahamic and Enlightenment presuppositions with ones more often associated with certain pre-Columbian American cultures, and their promotion of the education in "self-willed, self-wrought personality excellence" that John Collier ascribes to some Native American/First Nations societies.[96] This personality excellence, centered around a concept of personal honor defined as voluntary self-limitation, aims toward an application not often promoted in modern Western literature: rather than enabling individual achievement or glory, it instead seeks to hold the individual within the social matrix, which is ideally (in these novels' value systems) an extension of the ecological matrix.

In the process, *Tuck Everlasting* and *The Farthest Shore* shame and stigmatize a significant number of modern-day orthodoxies. These range from profiteering, land-development, and technological progress to a belief in the Chain of Being and immortality as an Edenic state. The messages deliver to chapter-book readers, in accessible, cleverly conceived, and well-crafted fictive form, the Deep Ecology tenets that Arne Naess would outline in his 1973 treatise "The Shallow and the Deep, Long-Range Ecology Movement": namely, that non-human life has intrinsic merit, regardless of its perceived or immediate usefulness or lack of usefulness to humanity; that human overpopulation is a depredation upon the world; that species diversity is crucial. In their argument for a deeply inculcated sense of self- and therefore species-limitation, both novels issue a credo approximating Naess's argument that "Humans have no right to reduce this richness and diversity except to satisfy vital human needs."[97]

Le Guin offered her proto–Deep Ecological platform two years before Naess published his, Babbitt two years after. This means that English-speaking child readers have had access to these ideas—recognized within the adult, non-fictional realm as subversive and countercultural—since they first coalesced in Western society. Nonetheless, not only did the phalanx of children's literature gatekeepers decline or fail to flag the books' heterodox content, but, as I mention at the top, they readily ushered both books into the heart of the children's canon.

Although the promises of unnatural life-extension to which the novels appear to be responding have in the past 40 years proven to be premature, if not false altogether, this brief, concentrated push against immortalist thinking appears to have helped widen the gateway into children's literature for a newly overt pro-ecological set of messages.

Notes

1 Michael R. Collings, "The Mechanics of Immortality," in *Immortal Engines: Life Extension and*
Immortality in Science Fiction and Fantasy, eds. George Slusser, Gary Westfahl, and Eric S. Rabkin (University of Georgia Press, 1996), 36.
2 Natalie Babbitt, *Tuck Everlasting* (Farrar, Straus and Giroux, 1991), 64.
3 E. O. Wilson, *Biophilia* (Harvard University Press, 1984), 118.
4 Wilson, *Biophilia*, 120.
5 Paul Lawrence Farber, *The Temptations of Evolutionary Ethics* (University of California Press, 1994), 20.
6 Ted Westhusing, "A Beguiling Military Virtue: Honor," *Journal of Military Ethics* 2, no. 3 (2003): 200.
7 Wilson, *Biophilia*, 120.
8 Abrash, "Is There Life After Immortality?", 19.
9 Abrash, "Is There Life After Immortality?", 19.
10 Samuel Delany, *The Jewel-Hinged Jaw: Notes on the Language of Science Fiction* (Wesleyan University Press, 2009). Qtd in Susan Bernardo and Graham

J. Murphy, *Ursula K. Le Guin: A Critical Companion* (Greenwood Press, 2006), 14.
11 Alan Harrington, *The Immortalist* (Random House, 1969), 3, 21–22.
12 Ursula Le Guin, *A Wizard of Earthsea* (Bantam, 1980), 6.
13 Le Guin, *Wizard of Earthsea*, 5.
14 Le Guin, *Wizard of Earthsea*, 18.
15 Ursula Le Guin, *The Farthest Shore* (Bantam, 1983), 133.
16 Le Guin, *Farthest Shore*, 156.
17 Le Guin, *Farthest Shore*, 21.
18 Le Guin, *Farthest Shore*, 38.
19 Babbitt, *Tuck Everlasting*, 37.
20 Babbitt, *Tuck Everlasting*, 27–29.
21 Babbitt, *Tuck Everlasting*, 7.
22 Babbitt, *Tuck Everlasting*, 63.
23 Babbitt, *Tuck Everlasting*, 85.
24 Babbitt, *Tuck Everlasting*, 86.
25 Babbitt, *Tuck Everlasting*, 8.
26 Le Guin, *Farthest Shore*, 75.
27 Le Guin, *Farthest Shore*, 178.
28 Le Guin, *Farthest Shore*, 53.
29 Le Guin, *Farthest Shore*, 8.
30 Le Guin, *Farthest Shore*, 96.
31 Le Guin, *Farthest Shore*, 99.
32 Le Guin, *Farthest Shore*, 121.
33 Le Guin, *Farthest Shore*, 136.
34 Le Guin, *Farthest Shore*, 136.
35 Babbitt, *Tuck Everlasting*, 6.
36 Babbitt, *Tuck Everlasting*, 15.
37 Babbitt, *Tuck Everlasting*, 111.
38 Babbitt, *Tuck Everlasting*, 14.
39 Babbitt, *Tuck Everlasting*, 112.
40 Babbitt, *Tuck Everlasting*, 50.
41 Babbitt, *Tuck Everlasting*, 23.
42 Babbitt, *Tuck Everlasting*, 25.
43 Babbitt, *Tuck Everlasting*, 45.
44 Babbitt, *Tuck Everlasting*, 63.
45 Babbitt, *Tuck Everlasting*, 138.
46 Farah Mendlesohn, *Rhetorics of Fantasy* (Wesleyan University Press, 2008), 115.
47 Mendlesohn, *Rhetorics*, xxii.
48 Mendlesohn, *Rhetorics*, xx.
49 Babbitt, *Tuck Everlasting*, 18.
50 Babbitt, *Tuck Everlasting*, 98.
51 Babbitt, *Tuck Everlasting*, 102.
52 Babbitt, *Tuck Everlasting*, 74.
53 Babbitt, *Tuck Everlasting*, 99.
54 Babbitt, *Tuck Everlasting*, 100.
55 Babbitt, *Tuck Everlasting*, 74.
56 Babbitt, *Tuck Everlasting*, 97–98.
57 Babbitt, *Tuck Everlasting*, 25–26.
58 Babbitt, *Tuck Everlasting*, 43.
59 Babbitt, *Tuck Everlasting*, 28.

60 Babbitt, *Tuck Everlasting*, 72.
61 Babbitt, *Tuck Everlasting*, 132.
62 Le Guin, *Farthest Shore*, 75.
63 Le Guin, *Farthest Shore*, 14.
64 Le Guin, *Farthest Shore*, 74.
65 Le Guin, *Farthest Shore*, 135.
66 Le Guin, *Farthest Shore*, 178.
67 Le Guin, *Farthest Shore*, 177.
68 Le Guin, *Farthest Shore*, 179.
69 Wilson, *Biophilia*, 120.
70 Le Guin, *Farthest Shore*, 178.
71 Le Guin, *Farthest Shore*, 160.
72 Le Guin, *Farthest Shore*, 181.
73 Babbitt, *Tuck Everlasting*, 41.
74 Babbitt, *Tuck Everlasting*, 61–62.
75 Babbitt, *Tuck Everlasting*, 63–64.
76 Lynn White, Jr., "The Historical Roots of Our Ecologic Crisis," in *The Ecocriticism Reader: Landmarks in Literary Ecology*, eds. Cheryll Glotfelty and Harold Fromm (University of Georgia Press 1996), 9.
77 Babbitt, *Tuck Everlasting*, 3.
78 Babbitt, *Tuck Everlasting*, 8.
79 Babbitt, *Tuck Everlasting*, 8.
80 Le Guin, *Wizard of Earthsea*, 17–18.
81 Bernardo and Murphy, *Le Guin Critical Companion*, 91.
82 Le Guin, *Farthest Shore*, 66.
83 Le Guin, *Farthest Shore*, 66.
84 Le Guin, *Farthest Shore*, 67.
85 Babbitt, *Tuck Everlasting*, 53.
86 Babbitt, *Tuck Everlasting*, 69.
87 Babbitt, *Tuck Everlasting*, 97.
88 Babbitt, *Tuck Everlasting*, 118.
89 Babbitt, *Tuck Everlasting*, 5.
90 Babbitt, *Tuck Everlasting*, 116.
91 Le Guin, *Farthest Shore*, 50.
92 Le Guin, *Farthest Shore*, 87.
93 Le Guin, *Farthest Shore*, 128.
94 Le Guin, *Farthest Shore*, 142.
95 Le Guin, *Farthest Shore*, 154.
96 John Collier, *Indians of the Americas: The Long Hope* (Mentor Books, [1947] 1960), 12.
97 Arne Naess, "The Shallow and the Deep, Long-Range Ecology Movement," in *Foundations of Environmental Philosophy*, ed. Frederik A. Kaufman (McGraw-Hill, 2002), 405.

Bibliography

Abrash, Merritt. "Is There life After Immortality?" In *Death and the Serpent: Immortality in Science Fiction and Fantasy*, edited by Carl Yoke and Donald M. Hassler. Greenwood Press, 1985.

Babbitt, Natalie. *Tuck Everlasting*. 1975. Sunburst/Farrar, Straus and Giroux, 1991.

Bernardo, Susan, and Graham J. Murphy, eds. *Ursula K. Le Guin: A Critical Companion*. Critical Companions to Popular Contemporary Writers series. Greenwood Press, 2006.

Collier, John. *Indians of the Americas*. 1947. Mentor, 1960.

Collings, Michael R. "The Mechanics of Immortality." In *Death and the Serpent*, edited by Carl Yoke and Donald M. Hassler. Greenwood Press, 1985.

Delaney, Samuel. *The Jewel-Hinged Jaw: Notes on the Language of Science Fiction*. Wesleyan University Press, 2009.

Farber, Paul Lawrence. *The Temptations of Evolutionary Ethics*. University of California Press, 1994.

Harrington, Alan. *The Immortalist*. Random House, 1969.

Le Guin, Ursula K. *The Farthest Shore*. 1972. Bantam, 1983.

Le Guin, Ursula K. *A Wizard of Earthsea*. 1968. Bantam, 1980.

Mendlesohn, Farah. *Rhetorics of Fantasy*. Wesleyan University Press, 2008.

Naess, Arne. "The Shallow and the Deep, Long-Range Ecology Movement." In *Foundations of Environmental Philosophy*, edited by Frederik A. Kaufman. McGraw-Hill, 2002.

Slusser, George, Gary Westfahl, and Eric S. Rabkin, eds. *Immortal Engines: Life Extension and Immortality in Science Fiction and Fantasy*. University of Georgia Press, 1996.

Vasbinder, Samuel H. "Deathless Humans in Horror Fiction." In *Death and the Serpent*, edited by Carl Yoke and Donald M. Hassler. Greenwood Press, 1985.

Westhusing, Ted. "A Beguiling Military Virtue: Honor." *Journal of Military Ethics* 2, no. 3 (2003): 195–212.

White, Lynn, Jr. "The Historical Roots of Our Ecologic Crisis." In *The Ecocriticism Reader: Landmarks in Literary Ecology*, edited by Cheryll Glotfelty and Harold Fromm. University of Georgia Press, 1996.

Wilson, E. O. *Biophilia*. Harvard University Press, 1984.

Yoke, Carl B. "Why Was I Born If It Wasn't Forever?" In *Death and the Serpent*, edited by Carl Yoke and Donald M. Hassler. Greenwood Press, 1985.

Yoke, Carl B., and Donald M. Hassler, eds. *Death and the Serpent: Immortality in Science Fiction and Fantasy*. Greenwood Press, 1985.

4 The children strike back

Anti-immortalism and social revolution in *The Amber Spyglass* and *The Other Wind*

A key part of my argument to this point has been that the preponderance of children's literature has persistently fostered and affirmed what E. O. Wilson's biophilia hypothesis calls our "innate affinity for life,"[1] our "ultimate obedien[ce] to biological law,"[2] and our experience of ourselves as "in the fullest sense a biological species [that] will find little ultimate meaning apart from the remainder of life,"[3] and that by doing so, the genre has for the past century or so acted as an overlooked site of resistance to modern Western society's divorce from nature. I have suggested that adults over the past 150 years have used the Romantic conception of the child, including the Rousseauean ideal of connection with nature, as a cover for latently radical and increasingly disjunctive allegiance to biophilic priorities. I further contend that Western society as an organismic whole has used the relative cultural insignificance of children's literature to smuggle these biophilic values and perceptions safely down the generations from the time of the Industrial Revolution onward.

Extending from this contention, I have argued that by helping to articulate and amplify in the child reader a species self-perception and set of values sharply at odds with the prevailing mores of adult society, the genre has contributed to a schism which I have termed "internal anomie." This disjuncture comes into full effect when the child who has been taught to honor their biophilic urges grows old enough to participate actively in—and is, indeed, compelled to enter into—the jobs, lifestyles, and overarching systems of modern Western society. At this point, most adults in the Western and Westernized parts of the world find little place for the deep engagement with the natural world that is their evolutionary imperative, or for the conscious respect for it that the biospiritual culture of children's literature may have inculcated in them. Material survival usually demands that they directly or indirectly contribute to the degradation of the planet's biosystems, the dire implications of which only the most wilfully ignorant can nowadays remain unaware. Social survival demands that they subordinate their nature-focused interests and priorities to human-focused ones,

DOI: 10.4324/9781003784210-5

with little overlap between the two sets in the age of urbanization, techno-industrial capitalism, and a complex of cultural beliefs—whether stemming from religion, humanism, or the positivist sciences—that promote human domination over nature.

As with most moves between cultures with conflicting values, the young person living in the Western hemisphere (or in a westernized society) who moves from children's to adult society must live with irreconcilable sets of values and priorities. In the case of this particular cultural transition, they must cope also with a conflict between two of their deepest species urges: the social and the biophilic.

Looking at the past hundred-odd years of the genre, from "Golden Age" texts through a cluster of ecology-themed and anti-immortalist texts in the late 1960s and early '70s to the current-day trend of dystopic and post-apocalyptic novels, we can see an increasing tension between the genre's "Trojan Horse" role and its potential as a vanguard cultural weapon in the gathering ecological resistance movement. The "smuggling" role required the heterodox biotheism of children's literature to be presented in inconspicuous ways. The vanguard role requires it to declare its contents openly.

Books of the 19th-century fin de siècle and the end of the Victorian era, such as *The Secret Garden* and *The Wind in the Willows*, act as gentle, though important, agents of subversion, stimulating a Counter-Enlightenment allegiance to our animalistic, evolutionary sense of identity. Toward the middle of the 20th century, as I have observed in previous chapters, a more urgent and overt rhetoric emerges that seeks to inculcate in its readers a strong sense of the self-limiting trait of honor and a concomitant rejection of the unnatural, unbiological species perception represented by immortalism.

With Ursula K. Le Guin's *The Other Wind* (2001) and Philip Pullman's *The Amber Spyglass* (2000), the internalized conflict comes snarling out into the open. These millennial novels, along with the series to which they belong—respectively, Le Guin's Earthsea cycle and Pullman's His Dark Materials trilogy—make an annunciatory call to arms. Both books depict child-led insurrections against the spiritual paradigms and sociopolitical systems of their fictional worlds, with their goal being to overturn the Western dispensation entirely and start afresh. Thirty years on from the hopeful spate of environmentalist texts that characterized children's fantasy of the late 1960s and 1970s, and 100 years after syncretic nature-worship infiltrated children's literature, these books forgo the gentle counsel of the child reader and cast their characters into the consequences of those earlier counsels having proven insufficient. They reflect the frustration and desperation, perhaps even despair, of what Al Gore has called the "impasse" of living in a time of global environmental damage, when "the maximum that is politically feasible...still falls short of the minimum that is scientifically and politically necessary" to avert environmental catastrophe.[4]

Both series feature young revolutionaries: Pullman's the street-smart urchin Lyra Belacqua and ruthless Will Parry, Le Guin's the half-burned rape survivor Tehanu and exploited village girl Irian. These rebels fight long-entrenched power structures, in both cases dominated by privileged majority-race men, that have given generation after generation of children no choice but to enter into a hubristic, short-sighted, nature-spurning way of life directly in conflict with their biophilic instincts and values.

Tehanu, the titular heroine of the fourth novel in the Earthsea cycle, in *The Other Wind* is the reluctant and largely silent undoer of Earthsea's exclusively male hierarchy of magic-working. One of the few born to each generation of humans who is also a dragon—or, as the dragon Kalessin notes, perhaps a dragon who is also human—Tehanu is the messenger and herald of choice for the systemically arrogant West. She plays a key role in events that compel Earthsea's most powerful men to consider the possibility that the Western culture's decision to separate from the rest of the natural world in order to pursue material wealth, possessions, and technological knowledge was "a bad bargain from the beginning"[5] and helps to catalyze a socio-spiritual revolution that forces the imperious mages of Earthsea to abjure their ability to summon dead souls, eradicate their land of the dead, and "open [their] doors" to women,[6] with their different ways of knowing and different modes of power.

Using her dragon-ability to travel to the land of the dead and her dragon-strength to lift the leaden stones of the immovable wall enclosing the world of the dead, Tehanu, along with her dragon-woman sister Irian and some of the humbler mages of Roke, dismantles the mage-made boundaries to let the corralled souls atomize into a completed death.

Pullman's Lyra first helps her father, the rogue theologian Lord Asriel, in his campaign to overthrow the Authority (i.e., the Church), the conformist angels, and the God-Regent Metatron. Then, in line with what Arthur Bradley and Andrew Tate call Pullman's "open hostility to any force—human or divine—that aspires to omnipotence,"[7] she rejects this alliance, as she comes to realize that, given Asriel's hunger for limitless power, the new boss is much like the old boss. As much to the point, she learns that his arrogant ambition has led him to overlook the fact that people's daemons, the animal-formed part of their spirit, can live only in the worlds in which they were born, and thus his efforts to create a separate "Republic of Heaven" are, as Lyra exclaims, "[A]ll wasted! All for nothing!"[8]

Lyra strikes out with her own ragtag company to fulfill her prophecy as a "second Eve"[9] who will not only reverse the conventional negative implications of the "Fall" into consciousness and sexual knowledge, but also be the means by which "Death is going to die."[10] She achieves this by taking her companion and eventual lover, Will, to the underworld, where they open a window to let the dead souls dissolve to "be alive again in a thousand blades of grass, and a million leaves...be falling in the raindrops and blowing in a fresh breeze ...be glittering in the dew under the stars and the moon out there in the physical world, which is our true home and always was."[11]

Lyra and Will complete the revolution by making sure that the windows between the worlds are closed in order to restore all the worlds' ecosystems to intactness, and then breaking the "subtle knife" that has been used to create the windows. Lyra's final significant act in Pullman's His Dark Materials trilogy is to accept that she and Will must live in separate worlds so that the windows between worlds can be closed and their essential substances once again contained within them. Sarah Cantrell gives this a convincing ecocritical reading as Lyra and Will learning "that limits are the prerequisites for restoring fragile universes."[12]

In depicting revolutions that reconstruct the Western world from the beginning, seeking out and deeply revising the aspects of founding mythologies that have put Western culture out of a "right relationship" with nature and set us on a trajectory for disaster, *The Other Wind* and *The Amber Spyglass* occupy a position both within and above the body of dystopian and post-apocalyptic writing that Carter F. Hanson considers "the default setting" for imagined societies in current youth fiction. They present what Kay Sambell calls "admonitory scenarios" whose "primary purpose is to puncture old myths and dreams [about]...what human aspirations and ideals are *really* likely to mean for the future of mankind" and "violently explode the myth that science and technology will bring about human 'progress'."[13]

However, even while both texts thereby join the intense current discussion about posthumanism, they refute several prevalent notions about posthumanist literature for young people: that it typically "seek[s] to protect the liberal humanist subject,"[14] helps young people "define themselves as human ... using the standards of morality set by the liberal humanist model,"[15] and in an overall way shows a "reluctance... to abandon liberal humanism."[16] Granted, it is perhaps valid to read a defense of humanism in those children's and YA novels that focus on human subjects in isolation, such as Lois Lowry's *The Giver*, Marilyn Kaye's Replica series, and Frank Bonham's *Forever Formula*, since the threat they depict is directly to humans and thus may come across as defending an orthodox human self-perception only. As well, there is no denying that Philip Pullman is conspicuously unable to pull free of humanism's gravitational pull: he tacks a Victorian-style moral ending onto his trilogy that has dismayed many admirers even while it placates Christian critics.

That said, the strong and longstanding tradition of children's and YA novels that fiercely resist infringing on the evolution-informed identities of other animals or *any* element of nature—from Joy Adamson's *Born Free* and Glendon Swarthout's *Bless the Beasts and the Children* to Sterling North's *Rascal*, Ken Oppel's *Half-Brother*, and Peter Dickinson's *Eva*—indicates instead that it is nature, and our identity as animals within nature, that this purportedly "new" trend is defending. Moreover, it does so in a way that runs explicitly counter to the human-exceptionalist premise of liberal humanism. In presenting what Helena Feder (following N. Katherine Hayles) terms "critical posthumanism"—that is, a "critique which reveals that the human of humanism…is itself a fantasy"[17]—*The Other Wind* and *The Amber Spyglass* turn a spotlight on the idea that "far from finding ourselves on the far side of a historic rupture, we may always have been posthuman."[18]

Le Guin's "second Earthsea trilogy" (to use Sandra J. Lindow and Darko Suvin's term for the belated additions *Tehanu*, *Tales from Earthsea*, and *The Other Wind*) and Pullman's His Dark Materials show these latest "trends" in publishing for young people to be part of a century-long tradition of representing, promoting, and defending our identity as animals with rightful places in both evolution and the natural world. His Dark Materials (in particular *The Amber Spyglass*) and Le Guin's latter Earthsea books (in particular *The Other Wind*) are therefore useful not only in representing current trends as a whole but also in presenting a wide enough scope, with their re-examinations of whole mythopoeias, to highlight the persistence of biophilic, evolutionary, and biotheistic vision in the history of children's literature publishing.

Although I say that the differences between these books and those in earlier chapters are of degree rather than kind, overtness wreaks its own changes. Thus, my initial revisiting of the elements of biophilia, biotheism, and "internal anomie," centered on conceptions of immortality, finds each of these elements transformed by the current acceptance of youth literature as a radical, even militant genre. The first element I examine in this chapter, the novels' stand on immortalism, is representative of this social permission for militancy. In the wholesale toppling of world orders (or at least Western sociocultural paradigms) that both series call for and depict, the attacks on systemic immortalism are the most salient acts, both materially and symbolically. Le Guin and Pullman use Western humanity's "interrupted" deaths—in both cases created by the male ruling class and interwoven into creation myths as an unquestioned eternality—to symbolize a disregard for the rest of the living world that advantages the powerful elite (the mages of Earthsea, the Authority/God of Lyra's world) while causing powerful disturbances to the ontological whole.

At the same time, as I go on to argue, the rhetorical battle over immortalist thinking reaches an even greater explicitness than in first-wave anti-immortalist children's novels such as *Tuck Everlasting* and *The Farthest*

Shore. The negative implications of immortality in *The Amber Spyglass* and *The Other Wind* result from a clear anti-immortalist thrust that sees the pursuit of eternal life as the ultimate symptom of man's/men's catastrophic arrogance. Both Le Guin and Pullman put the issue of eternal life into protracted dialogue among their protagonists, proposing that the true antithesis to life is changelessness, not death, and that death itself is a point of joyous reunion with the rest of nature.

The trope of honor, which I revisit next, has also undergone a change. In earlier chapters I contend that honor emerged in ecocentric children's fantasy with the potential for life extension technologies in the 1960s and '70s, as a central mode of resistance to immortalism. Thirty or forty years on in the history of children's publishing, the notion of honor is still conspicuously present, but in *The Other Wind* and *The Amber Spyglass* it has been qualified in a way commensurate with not only the new urgency of the message but the newly open radicalism of youth literature. Although Le Guin and Pullman continue to foreground the importance of self-limitation and self-restraint, they seem to have concluded that for the pro-nature side of this battle, consistently exercising honor is a handicap it cannot afford. Where members of the power structure might take advantage of their opponents' honor to maintain the status quo, sympathetic characters in these stories are allowed to act on power advantages without authorial censure—and even, in some cases, with approval.

Following this, I show the Romantic conception of the child—which I have from the start perceived as a front for the cultural preservation and transmission of biophilic values—to be largely overturned in these books. While critics such as Kay Sambell and Kristine Moruzi see a continued tendency of millennial-dystopia authors to protect the Romantic child, and see "any childhood 'fall from grace'" as "caus[ing] a dilemma for children's authors, as they struggle to avoid valorizing the values of the polluted society,"[19] Pullman and Le Guin show little sign of such a struggle. Instead, to use Sarah Cantrell's words, the presence in these series of both sympathetic and antagonistic children "confirms the argument that childhood is not the preserve (or space apart) of romantic notions of innocence, kindness, pity and compassion. Rather, childhood is no less complicated nor less ethically questionable than adulthood."[20] By the same token, adults in these novels are no longer on the other side of a barrier that prevents them from showing allegiance to a pro-nature, biotheistic revolution. Both the latter books in the Earthsea cycle and the His Dark Materials trilogy feature ensemble protagonists, made up of children and adults, the powerful and the humble, insiders and outcasts, humans and non-humans (Gallevespians, witches, polar bears, dragons, domestic cats, trees, angels).

Within these ensembles, girls, women, and other female characters play the lead part. I finish the chapter by analyzing the significant new presence of ecofeminism in both books/series. I see this presence as stemming jointly from two factors. The first is a female-centered trend in marketing and publishing, with the phenomenal success of Suzanne Collins' Hunger Games heroine Katniss Everdeen probably being the inciting incident in this regard. The second is a growing perception that the originating point of Western society's disastrous existential misalignment is what Karen J. Warren calls "the logic of domination"[21] that led the male elite to establish in Western and Westernized lands not only fear- and hubris-driven paradigms of false immortality, but also antagonistic hierarchies between humans and nature, men and women, and different human cultures. Both Pullman's and Le Guin's series offer fictional explorations of ecofeminist Val Plumwood's assertion, "Forms of oppression from both the present and the past have left their traces in western culture as a network of dualisms, and the logical structure of dualism forms a major basis for the connection between forms of oppression."[22]

The issue of feminism is one of the many which lack of space obliges me to give short shrift. Given that *The Other Wind* and *The Amber Spyglass* belong to long, complex series, and that in each case those series have their own equally complex backstories (in Le Guin's case, the first trilogy of the Earthsea cycle; in Pullman's, the originating texts of John Milton's *Paradise Lost* and William Blake's *The Marriage of Heaven and Hell*, among others), I have had to skim over some elements and leave out others altogether. For instance, Pullman's intriguing conception of the "daemon" deserves to be analyzed from a number of ecocritical angles, to supplement Maud Hines' partly ecocritical reading and the numerous analyses that position the daemon within religious, philosophical, and psychological discussions. Similarly, the relationship between women and dragons in the second Earthsea trilogy, and how it figures in Le Guin's "revisioning" of Earthsea from the "pseudo-genderless male view point of the heroic tradition" to a militantly feminist perspective,[23] could sustain a book in itself—and hopefully will. Fortunately, much of this territory has been covered by other critics, not least among them Philip Pullman and Ursula K. Le Guin themselves, as these two authors are indefatigable commentators upon their own texts and adept literary critics in their own right.

Immortalism as cause and symptom

It should be said at the outset that neither Pullman's nor Le Guin's works are explicitly environmentalist or even nature-focused. Each has an overt preoccupation other than nature-human relationships. Pullman focuses chiefly on the damage wrought by the Church, and on a strategy to "end ... all those centuries of darkness!"[24] In theory, this could be an inherently ecocentric move, if one accepts the premises that Judeo-Christianity is a largely anti-nature belief system—one that, as Lynn White Jr. puts it, "by destroying pagan animism ... made it possible to exploit nature in a mood of indifference to the feelings of natural objects"[25]—and that "modern technology is at least partly to be explained as an Occidental, voluntarist realization of ... [Judeo-Christianity's] dogma of man's transcendence of, and rightful mastery over, nature."[26] However, as numerous critics have noted, the trilogy reinforces Christian perspectives nearly as much as it subverts them, and hence stops well short of moving to a genuinely ecological position. Richard Gooding believes that Pullman's "anxiety-laden" texts reflect the author's internal conflict regarding the logical outcomes of his "critique of the liberal humanist subject";[27] other critics argue that by virtue of writing a heavily intertextual counterargument to Christian doctrine and therefore "[lighting] his fire from the Christian flame,"[28] Pullman largely remains "of God's party"[29] and is "explicitly dedicated to critiquing the Christian meta-narrative from within."[30]

Whatever the source of the equivocation, there is no denying that Pullman peripheralizes environmental issues in the early going, and that he also works against some primary principles of ecological/environmental stances. For instance, he contradicts the Deep Ecology premise that all species have equal value and are interdependent with his open privileging of humans and the human-like sentient beings the *mulefa*. When the scientist Mary Malone fashions a glass through which she can see the mysterious and precious subatomic substance variously called Dust, *sraf*, and Shadows, she sees that "anything associated with human workmanship and human thought was surrounded by shadows."[31] Such passages elevate humanity in a way that also reinforces humanist assumptions about our species' monopoly on "consciousness," as well as the self-arrogating premium put on the trait. Pullman scholar Millicent Lenz reads the trilogy as affirming the Biblical idea that "[The] wresting of consciousness from nature results in guilt, because it breaks the 'bond' with earth...but only by sundering the bond [that] ...kept the original parents unaware, unthinking, merely intuitive like the animals could humanity reach intellectual and moral maturity."[32] Certainly the series contains abundant textual support for this type of ultra-humanist reading, suggesting that the environmental focus that enters later sections of the trilogy is largely incidental.

Le Guin's "second Earthsea trilogy" also has a different main focus: the righting of an imbalance that expresses itself mostly as a female-excluding power structure and the resultant dearth of women's wisdom in the operations of the land. *Tehanu*, in particular, has received much critical attention as "a fictional experiment in the feminist appropriation and subversion of the hero-tale."[33] As Sandra J. Lindow reads it, all three books—*Tehanu*, *Tales [from Earthsea]*, and *The Other Wind*—"see[m] to be consciously remythologizing Earthsea in a way that balances feminine and masculine principles."[34] Indeed, Le Guin herself has laid out her "revisioning" of Earthsea as deliberately "refocus[ing] the narrative perspective to show the reader the world 'as seen through a woman's eyes'," in her 1993 chapbook *Earthsea Revisioned*.[35]

The emphasis on the inherent value of all existing entities that Le Guin shows in the first three books of the Earthsea cycle she here subordinates to a change in gender dynamic and power dynamics within society. As well, she adopts something of a resigned dualism, accepting the self-willed separateness of humans from other animals, although her conception of the "Vedurnan"—the pact of division between humans and dragons—carries more than a few overtones of the Expulsion from Eden. The mages gathered on Roke describe the Verdurnan as "the choice we made"; "the knowledge of good and evil"; "The joy of making, shaping…Our mastery"; "And our greed, our weakness, our fear."[36] Even while sadly describing this version of the "Fall," Le Guin, like Pullman, arrogates the human species, having even the Earth-worshipping Kargish people of Earthsea's eastern regions believe that people and dragons are the incarnations to aspire to, as they are the sacred beings. For a shot at this ultimate rebirth, they observe foundational precepts, avoid taboos, and strive to be good. Warren G. Rochelle, who considers Le Guin "[Ralph Waldo] Emerson's spiritual heir, as much as are Thoreau, Fuller, Whitman and C.S. Peirce,"[37] believes that within the Earthsea texts Le Guin is using this dualism to represent a conscious human-species "Emersonian choice" between being and doing, with "doing" coming along with a knowledge of good and evil: to wit, "dragons chose to be Nature; humans chose to be a part of nature."[38]

Yet both *The Amber Spyglass* and *The Other Wind* are framed by, and increasingly encroached upon by, environmentalist concerns, and both series move almost as if compelled from social dystopias or dysfunctions to a broader vision of "awryness" that has to do with humanity's fundamental conceptions and practices of its relationship with the rest of nature. In Pullman's His Dark Materials trilogy, windows have been cut and holes rent between worlds, piercing universal ecologies, causing massive biospheric disturbances, and letting in invasive agents. The holes also let out the Dust that symbolizes not only individual consciousness, wisdom, and self-awareness, but also the integrity and particularity of a place—in a manner of speaking, any given world's "daemon," in the sense of its interdependence and inter-identity with the beings making up that place.

In Lyra's world, the polar bears' Arctic habitat is warmed and flooded by the sunlight from a neighboring world. In another world that she and Will journey to, an abyss between worlds is destroying the Dust-to-tree relationship that sustains the world's ecosystems, including the sentient beings who live there. As Sarah Cantrell reads it, "By setting his narrative among multiple universes, Pullman ...challenges readers' notions of the scale and scope of environmental dystopia"; in her view, he is engaging in the "radical trend in children's texts [of] exposing readers to environmental catastrophe."[39]

In *The Other Wind*, global disturbance is also taking place, though in the more oblique and symbolic manner befitting a fantasy series in which magic stands in for both technology and religion. The breach of the wall between life and death made by the immortalist mage Cob and healed by Archmage Ged and his companion Prince Lebannen, as narrated in the final book of the initial trilogy, *The Farthest Shore*, has made the wise of Earthsea aware of a systemic misalignment woven historically into the social structures of all Western Earthsea. The three main harbingers of deep change are, firstly, the dragons breaking their oath to stay away from the western islands and beginning to lay waste to human habitation; secondly, the inability of Roke's master mages to find a new Archmage; and, thirdly, a new clamoring among the dead souls of the western islands for release from their false immortality. The wizard Onyx tells other mages gathered at Roke, "We sit here talking on the edge of a precipice—we all know it."[40]

Although Le Guin describes the consequences of the current dispensation continuing mostly in individual and human terms, rather than ecological ones, the dragons, deep, wise, and wild, act as representatives of nature pushed beyond the breaking point by the West's system of exploitation and dishonor. The dragons "signal change, the coming of the destruction of [what Le Guin calls] the 'artificial order of oppression'."[41] Along with the dragons, the whole system rebels, though subtly. The Master Patterner of Roke says, "[A]ll I can hear the leaves say is Change, change ...Everything will change but them."[42] Darko Suvin describes the difference between the "first" and "second" Earthsea trilogies as showing that although "the world can perhaps be saved from one acute danger by a feat of heroism...the pervasive system of cruel power and privilege cannot be righted by one explosive outburst...but only by protracted and complex collaboration for life between women, men, and dragons (and animals and plants)."[43]

Thus, *The Amber Spyglass* and *The Other Wind* might on the face of it seem to fall into Stephanie Guerra's category of "novels that are not primarily focused on environmental issues ...[but] include ecological damage...as a subtext."[44] I would argue, though, that they do more than this, because the climaxes of the books are not about the Church, per se, or the triumph of feminism, but instead the release of the penned-up dead souls of Western inhabitants from misbegotten, unnatural afterlives. These afterlives, in both sets of texts, take place in a dry, dim, Hades-like land of the dead, full of milling souls tormented by changelessness (and, in Pullman, sleep-deprivation). These souls long to finish dying—"To be one with the earth again. To rejoin it," as *The Other Wind* puts it;[45] to be "part of everything alive again," as *The Amber Spyglass* says in remarkably similar language.[46]

As Graham Holderness observes, the release of souls within Pullman criticism has been "undervalued by [the] stress on the Gnostic structure of the trilogy" and "tends to be read merely as an episode in the overall denouement of the trilogy…not perceived as the main achievement of the novels."[47] I agree with Holderness that in fact the episode "is both the narrative climax and the most fully realized imaginative achievement of the novels."[48] Meanwhile, the "second Earthsea trilogy" has so commonly been read as "a womanist version of the first trilogy" that it has been overlooked as (in Darko Suvin's terminology) "a dragonist revision,"[49] in the sense that dragons in the latter Earthsea books represent "the nature of things, of the badly made human contract with the cosmos …rebelling against the contractors."[50]

What does this shared narrative turn signify? Unlike plot elements such as the overthrow of the church or new roles within the power structure for women, the demise or neutralization of a land of the dead has no real-life correlative: it cannot actually occur, since these lands and the ghosts/dead souls in them are fantastical. What, then, are Philip Pullman and Ursula K. Le Guin trying to say in making this overthrow the pinnacle of action, and indeed the goal of their heroes' quests?

One plausible interpretation is that these tormenting, artificially created afterlives symbolize the psychospiritual break that Westerners have suffered as a result of the Christian belief system; of the Western technoscientific paradigm that was, as Lynn White contends, "cast in a matrix of Christian theology";[51] and of the "religion of humanity" that constitutes the humanist tradition.[52] By making the end of immortality the chief objective and supreme achievement of Lyra, Tehanu, and the other protagonists, the authors depict the recent religion- and science-based quest for unnatural long life, and the West's predication of human superiority upon this, as the point where human and nature became segregated in the view of Western society—a segregation that in turn acts as the root cause of social, psychological, and environmental catastrophe. In other words, Pullman and Le Guin point the finger specifically at immortalist thinking, both faith-based and ostensibly secular—and therefore both post-death and pre-death—as the master narrative in the "logic of domination." Their shared indictment of the way the Western men in positions of power desire "to own life, possess it, as if it were a jewel in a box," and their craving to live forever,[53] forms both trilogies' most directly ecological message, in being the aspect most concerned with the question of humans' unique and non-unique position within and without nature. By virtue of being the most foregrounded element of the plot, despite not being key elements of the theme, this fundamentally ecological concern becomes obliquely centralized.

Both *The Amber Spyglass*'s land of the dead (complete with equally hellish "suburbs") and *The Other Wind*'s "dry land" were created by powerful men who manipulated founding myths and treaties to demonstrate their ultimate mastery: that over death. In Le Guin, these men were the Rune-makers, who, envying the dragons' ability to fly on "the other wind," outside of time and body, sectioned off part of the dragons' realm with a wall that nothing living—not humans nor even dragons—can cross. Then, with their immense powers of naming and spell-casting, the Rune-makers laid a net of spells across Western Earthsea, so that when its people died, they would come to this "west beyond the west"[54] and "live" (or, more accurately, be trapped) there forever as spirits. The powerful mages ever since have fostered a chauvinism about their secular mageocracy. They scorn as "barbarians" the eastern peoples, who they deem to have forsworn their immortality by refusing to stop believing in the Old Powers of the Earth and learn the True Speech.

In Pullman, it is unclear whether the land of the dead was created by "the Authority" or is a natural waystation to which all people must go after death and before atomization. Either way, the Authority has since time immemorial governed the land and manipulated its once and future inhabitants, feeding the Western population the lie that, in the words of a long-disillusioned dead martyr, "when we died, we'd go to Heaven. And... Heaven was a place of joy and glory and we would spend eternity in the company of saints and angels praising the Almighty."[55]

Thus Western society as a whole is shown as suffering from a systemic immortalism that was initiated in the early days of male-dominated hierarchies, and the rupture with nature is shown as an effect of the same cause as religious oppression and gender oppression. All of these extend from what Val Plumwood terms a "master model"[56] with the ability, quite literally, to "direct, cast and script this ruling drama."[57] Both Pullman and Le Guin seem to perceive the compulsion to pursue immortality as a chief failing, almost an inherent vice, of those who have power, and an extension of power over nature that men of power will attempt to achieve, whether secularly or through religious doctrine.

When the lands of the dead emerge in these books—in *The Other Wind* from the start, in *The Amber Spyglass* not until the middle of the volume—other issues and plotlines are subordinated to them. It is in this subordination that both Pullman and Le Guin suggest that they recognize human-nature divorce as a problem of greater magnitude. Although Lyra's stated importance up to the chapters in *The Amber Spyglass* that deal with the land of the dead is that she is in some unexplained way "a second Eve, mother of all! Eve, again!,"[58] in the end her great legacy is not that she is able to become conscious, knowledgeable, and sexual without shame, but that she is able to persuade the harpies of the underworld to show her and Will a place to cut open a window to another world. She then persuades the (mainly eager) ghosts to walk through and let their individual identities dissolve into a different type of immortalism: one in which, as Lyra rhapsodizes to them, "all the particles that make you will loosen up and float apart" and they will become "part of everything. All the atoms...gone into the air and the wind and trees and the earth and all the living things."[59]

Underscoring the primacy of this segment in Pullman's envisaged revolution, and thus of the linchpin importance of immortalism to the system that his protagonists are overthrowing, Lyra's helpmeet Tialys says of Will's plan to release the ghosts, "This will undo everything. It's the greatest blow you could strike. The Authority will be powerless after this."[60] Lyra's father, Lord Asriel, echoes Tialys's perception that the end of immortalist manipulations equates to the end of the world order as it has been known throughout the Christian age, seeing even his own grand ambitions and accomplishments as merely the means to this end: He acknowledges that the sole purpose of creating the Republic of Heaven may have been to help Lyra end the prison of forced immortality and release the trapped souls into reunion with the rest of the universe.

Similarly, Le Guin re-values the great heroes and heroines of Earthsea according to how they contribute to helping make sure that the land of the dead is dissolved and, more importantly, what they represent about the potential for immortalist thinking and actions being eradicated or at least minimized in the new way of the world. King Lebannen, whom Le Guin had characterized conventionally in *The Farthest Shore* in the "mode of the Techno-Heroic" in which the conflict-worshiping, glory-mongering hero is put on "a stage or a pedestal or a pinnacle,"[61] now is deemed not the apex of Hardic (Western Earthsea) history, but rather, as the former archmage Ged muses, as "Only a beginning. A doorway...and he the doorkeeper. Not to pass through [ellipses in the original]"[62] His importance is in accepting that the wizards' interference with the laws of life and death endangers the entire cosmology. In contrast with *The Farthest Shore*, Lebannen's acid test in this novel is not the heroic journey across the seas in the company of the greatest mage the land has ever seen: it is to face his fear of the strong-willed Kargish princess Seserakh, whose people have never joined in with the western belief in the afterlife or sought to dominate nature by learning the Speech of the Making, but who still worship the Old Powers of the Earth and believe in reincarnation. They remember, as a wizard from the dark land of Paln puts it, "what the earth is, what life everlasting is."[63]

Likewise, Tenar, the heroine of the second Earthsea book, the Newbery Award-winning *The Tombs of Atuan*, is now mainly important not for her powers as a Kargish priestess, nor even as the first lover and only mate of the newly sexual and humanized Ged, but for her wisdom in seeing a middle reality between the technocracy of the West and the superstition of the East. Tenar tells a gathering of powerful mages and rulers, "I no longer believe that I am or was, as they told me…a single soul reincarnated endlessly and so immortal. I do believe that when I die I will, like any mortal being, rejoin the greater being of the world. Like the grass, the trees, the animals. Men are only animals that speak."[64] Meanwhile, the dragon-woman Irian, who has some months earlier destroyed the corrupted Roke mage Thorian, is now recognized to be the messenger or harbinger of a great event.

Where *Tehanu* is given over to mages, *The Other Wind* connects that "logic of domination" to the false immortality. Where in *The Farthest Shore* Le Guin characterizes the urge toward immortalism as a cause of immediate and catastrophic harm (specifically, the blighting of all creation, resulting from the mage Cob's misuse of the immortalist Lore of Paln), in *The Other Wind* she casts it as a symptom, too, of a complex of connected traits that have been socially systematized: men's power-hunger, need for control, arrogance, and fear-based rejection of "the powers of the Earth our mother."[65] Where *The Farthest Shore* envisages the impact on the earth and its ecosystems of Western cultures' nature-rejecting dualism, *The Other Wind* considers the reverse: the impact of dualism on the humans living within this worldview, who are psychically severed from the cycle of life and death—who are, in essence, forced to live a life in repudiation of their evolutionary and biophilic instincts. As the Pelnish wizard Seppel says, "Maybe it's not only our desire to live forever that has kept the wound [in the world] open, …but the desire of the dead to die."[66]

Exemplary lives and joyful deaths

As I say in the opening of this chapter, Pullman and Le Guin's texts generally propose solutions to the problems they identify, or at least present effectual responses to them. One common response they show to the immortalist indoctrination of the modern West, both programmatic and osmotic, is to not only cast as villains those who want to live forever, but to forcefully and repeatedly portray "completed" death as joyous, beautiful, and natural—in Darko Suvin's words, "a kind of Nirvana, in [a] very unorthodox post-Daoist cosmic syncretism."[67] In this re-education, Le Guin and Pullman effectively reverse the campaign of fear, portraying the eternal life of Christianity and secularity as horror-filled, and, by contrast, decomposition into the soil and dissolution into the air as fulfilling and blissful.

Le Guin presents this horror partly through nightmares that Tenar has about people with birds' wings and raptors' heads, their eyes circled with white and red. These stand or squat, unmoving, and make the air "so foul that she struggled to turn, to run away, but she could not move."[68] After waking, she recognizes that the dreams were of the paintings on the wall in her room in Havnor that have come to life—"Only it was not life. It was the endless, timeless unlife of those who died without rebirth: those accursed by the Nameless Ones: infidels, westerners, sorcerers."[69] In her old Kargish belief system, this is the fate inflicted upon westerners, who have been cast out by the Old Powers and the earth itself: they do not rejoin the living world, but go to a place of "half being" where, "winged but flightless, neither bird nor human, they must endure without hope."[70] In other words, in Tenar's Kargish culture, the cold, dim, unchanging land that the inhabitants of the West have long accepted as inevitable and eternal, the "afterlife" that Ged witnessed at the end of *The Farthest Shore* and has told Tenar about, is used as a cautionary tale against repudiating the powers of Earth, and told as a bogey-man story to young Kargs. In both Le Guin's and Pullman's books, dying and decomposing is a *reward* for the right attitude to life.

In Pullman's Limbo-like afterworld, the Authority has licensed flocks of harpies to "see the worst in every [ghost], and [feed] on the worst."[71] They wait until the ghosts are resting and then whisper to them all the bad things they did while they were alive. Worst of all, in both depictions, is the changelessness that here is shown as the true antithesis of life. One ghost says to Lyra and Will, "There ain't *nothing* going to happen! Nothing, forever!"[72] Lyra's child friend Roger, killed by Lord Asriel so that his life-force can be harvested to build a bridge between worlds, tells her, "This is a terrible place, Lyra, it's hopeless, there's no change when you're dead."[73]

By contrast, organic death is described in sometimes rapturous ways. Lyra tells the ghosts,

> This is what will happen...and it's true, perfectly true. When you go out of here, all the particles that make up will loosen and float apart, just like your daemons did...[T]hey're part of everything. All the atoms that were them, they've gone into the air and the wind and the trees and the earth and all the living things. They'll never vanish. They're just part of everything...You'll drift apart, it's true, but you'll be out in the open, part of everything alive again.[74]

When Will does finally cut a window to another world (as it happens, the world of the *mulefa*), it is described as the "sweetest thing" the ghosts have ever seen. As they breathe in the night air and look at the stars in the night sky, they are filled with hope and joy. "The first ghost to leave the world...laughed in surprise as he found himself turning into the night, the starlight, the air...and then he was gone, leaving behind...a vivid little burst of happiness."[75]

Le Guin's rhapsodic eternality is described through Roke Island's Immanent Grove, the forest that Segoy, the creator of Earthsea, planted and where "all things take their true nature."[76] The Grove models the ideal of immortality: it "is forever because it dies and dies and so lives."[77] Ged describes this immanence in dreamy terms, saying that the leaves glimmer, "like moonlight or starlight," even on dark days, and that as far as a person can walk, they will never come to the end of the grove.[78] And, like Pullman's ghosts, Le Guin's long-trapped dead go eagerly into dissolution, "walking with unhurried certainty towards the fallen places in the wall: great multitudes of men and women, who as they came to the broken wall did not hesitate but stepped across it and were gone: a wisp of dust, a breath that shone an instant in the ever-brightening light."[79]

Thus, the most direct remedy to the ecological problem that the two authors propose is the mitigating of immortalist sentiment through a strong counter-rhetoric, recasting natural death and decomposition as a joyful fulfillment of being. Tehanu, expanding on her foster-mother Tenar's matter-of-fact conception of rejoining the living world after death, encapsulates this newly positive view of going gently when she is on the verge of leaving her human form and going as a dragon to fly on "the other wind":

> [W]hen I die, I can breathe back the breath that made me live. I can give back to the world all that I didn't do. All that I might have been and couldn't be. All the choices I didn't make. All the things I lost and spent and wasted. I can give them back to the world...That will be my gift back to the world that gave me the life I did live, the love I loved, the breath I breathed.[80]

Le Guin and Pullman's repeated expressions of death as harmonious and a way to realign with nature rebuke the horror of "losing" the individual identity, as summed up by Eric S. Rabkin: "The persistence of the individual is a fantasy, clearly, yet a productive fantasy without which we would have no sense of self, and hence without which the very notion of immortality would be reduced to mere persistence, a state not unlike that of a rock."[81] The set of assumptions and anthropocentric valuations behind such a statement characterize the Western attitude toward the natural cycle of life and death, as well as toward non-human biotic and abiotic existence.

Beyond honor

A rhetorical overturning of death's stigma forms only one part of the trilogies' manifold program for righting the relationships that have grown disastrously askew from the influence of powerful male hierarchies. Another is the licensing of the heroic/protagonist mode to meet the opposition on its own level by temporarily and judiciously using some of its tactics and amorality, to the end of defeating it. If, as I argue in the previous chapter, *The Farthest Shore* and *Tuck Everlasting* represent an optimistic indoctrination into honor-based anti-immortalism, *The Other Wind* and *The Amber Spyglass* present what we might think of as "post-honor" scenarios. Where Le Guin and Babbitt's novels of the 1970s educate the child reader in the wisdom of declining to act on a power advantage, these novels of the new millennium exhibit what Roderick McGillis points out in his study of *fin de siècle* children's literature as "both utopian and dystopian, leav[ing] the future open-ended even as it posits an end-game."[82] More accurately, they show their worlds after the systemic forces of dishonor have overcome individual honor and thus nullified the species self-restraint which that trait was meant to foster. They show their worlds after the society-wide failure of that education. In Kay Sambell's words, they lead readers "to the recognition of the dire need to question official structure and to expose the corruption of the present adult world that could plausibly lead to such bleak and intolerable futures."[83]

Looking at the development of this particular aspect of ecologically oriented children's fantasy, we can conclude that Pullman and Le Guin, at least, have decided that to insist on honor is to tie one hand behind our backs. So, while their texts advocate "the necessity of living within one's limits" in order to "institute an 'ethic of care,'"[84] and while they stop short of allowing their protagonists to perform the worst violations available, they do not extend the insistence on honor to refraining from sometimes dishonorable and ignoble action against the powers of the system, which greatly outmatch them. The antagonists of *The Amber Spyglass* and *The Other Wind*—the Authority, with its agent Mrs Coulter (also Lyra's mother), "always ambitious for power";[85] Lord Asriel, the anti-church zealot feverishly determined to "build a bridge and cross …the barrier between the universes" (and also Lyra's father);[86] and the mage inheritors of Earthsea's Rune-Makers, who coveted the dragons' immortality and therefore claimed the timeless "west beyond the west" in pursuit of an immortal self—must be humbled by women, children, and the powerless before an alternative way of relating to the rest of creation can be established.

In order to do this, the honorable disenfranchised must solve the problem of honor's inherent disadvantage, which the English judge Lord Bowen captured in a witty but profound verse commenting on Matthew 5:45: "The rain it raineth on the just /and also on the unjust fella, / but mostly on the just because / the unjust steals the just's umbrella." The solution, these books propose to different degrees, is to set aside honorable self-restraint long enough to usher in a new dispensation in which "the model of disconnection [which] remains deeply and fatally entrenched in modern conceptions of the human and of nature"[87] has been displaced by gentler, less hierarchical attitudes and systems. When this is done, and the half-life of immortality turned into a completed death (which, as I argue above, symbolizes a wide-scale cultural acceptance of organic death, decomposition, and humanity's identity as a part of the animal kingdom and "in the fullest sense a biological species"[88]), then those who understand the need for species self-restraint will return to their honorability and their humbler way of interacting with the earth. In short, the "just" may steal their own umbrellas back from the "unjust," and even steal the unjust's umbrellas outright, if they do it as the means to an end.

Lyra has two sources of power with which she can combat the powerful Mrs Coulter and Lord Asriel's competing ambitions to, respectively, extend the Church's reach and depose God and the Authority altogether. One is her innate ability to decipher, untrained, the complex compass-like instrument of symbols called an alethiometer. This "golden compass" allows her to discern when someone else is telling the truth, know where other people are, and to a considerable extent know the future. Like Will's unique ability to possess and wield the "subtle knife" that can cut passages between worlds, Lyra's possession of and facility with the alethiometer is a "gifting" trope found commonly in fantasy, from Arthur's ability to pull the sword out of the stone to Lord Gwydion's ability to wield the death-killing sword Dyrnwyn and Harry Potter's selection of the only magic wand that can thwart the undead villain Voldemort.

More unusual is Lyra's second source of power: her ability to lie. Given that honesty is one of the most common and important expressions of honorability, this is a significant departure from the character make-up of most obligatorily moral protagonists. Lyra is shown dissembling in the face of power almost from the start of *The Golden Compass*, when she tries to avoid revealing the whole truth to Lord Asriel.

> "Where do you play to get so dirty?"
> She looked at him suspiciously. She had the feeling that being on the roof was forbidden, though no one had actually said so. "In some of the old rooms," she said finally.
> "And where else?"
> "In the clay beds, sometimes."
> "And?"
> "Jericho? Port Meadow?"
> "Nowhere else?"
> "No."
> "You're a liar. I saw you on the roof only yesterday."[89]

This early exchange, though unsuccessful on Lyra's part, sets the parameters of her dishonesty. She may (and should, and does) use her powers of deception in relations with opponents, in order to protect herself and advance her own ends. Indeed, one can fairly consistently tell the antagonists from the protagonists in the trilogy by whom Lyra lies to and whom she tells the truth to. Her interests are shown to be noble ones, if at times self-serving and at other times misguided. Her name, as many critics have pointed out, is an aural anagram of "liar," and the interpretation has been that this is deliberate on Pullman's part, "suggest[ing] storytelling (Lyra, lyre, lyric), invention, deception, trickery and bluff are at the very heart of Lyra's character."[90] When, in the land of the dead, she goes into storytelling mode with the ghosts, spinning glamorous yarns about royalty and romance, the harpy attacks her, "screaming 'Liar! Liar! Liar!' And it sounded as if her voice were coming from everywhere, and the word echoed back from the great wall in the fog, muffled and changed, so that she seemed to be screaming Lyra's name, so that *Lyra* and *Liar* were one and the same thing."[91]

Pullman draws attention to his authorial sanction of Lyra's dishonorable duplicitousness through Lyra's foster-mother, the gyptian Ma Costa. Ma admires the near-shape-shifting powers that Lyra achieves with her ability to dissemble and praises her deceptiveness. When Lyra, hurt, responds with the whole-cloth lie "I en't never deceived anyone!", Ma Costa replies, "Can't you see I'm paying you a compliment?"[92] Similarly, Lyra's surrogate father, the armored bear Iorek Byrnison, admiringly dubs Lyra "Lyra Silvertongue" for deceiving his great foe Iofur Raknison despite the fact that the bears famously see through deceit in a way that humans can't and can (purportedly) never be tricked.

Throughout the series, Lyra is given license to use this "dishonorable" power to counter other deceivers and destroyers. "Breaking the rules," as Lauren Shohet comments, "allows Lyric [sic] to keep alive the emancipating rebellion against tradition."[93] She is not expected to operate "umbrellaless" while the dishonest, ambitious wielders of power take advantage of her more restrictive morality. In the most dangerous and extreme situations, Lyra wins or escapes by using this "vice." One of the most signal occasions is when she tells Mrs Coulter, whose deceptive nature she has long since discovered, and whose evil program of severing children from their daemons she abhors, that her disappearance resulted from the gyptians kidnapping her, rather than her escaping and fleeing, as really happened. Her daemon, Pantalaimon, thinks to her, "We're only safe as long as we pretend." Pan "play[s] the same game: fool them, fool them."[94] And as Lyra spins her tale of guests at Mrs Coulter's party kidnapping her,

> she [feels] a little strength flowing back. And now that she was doing something difficult and familiar and never predictable, namely lying, she felt a sort of mastery again. She had to be careful not to say anything

> obviously impossible; she to be vague in some places and invent plausible details in others; she had to be an artist, in short.[95]

Kiera Vaclavik points out that "In the circumstances in which [Lyra] finds herself, lying and pretense are simply unavoidable: her various alter-egos are no mere play-acting whims but strategies upon which her life depends."[96] Granted that this is largely true, there are numerous books where heroic characters remain steadfastly honest even in equally dire circumstances, with the very point of their survival and eventual triumph being to valorize this refusal to buckle morally. Indeed, the line between the heroic and the villainous is usually drawn precisely along the distinction between truthfulness and duplicity. I would argue that the guiding principle behind the authorial approval of—even reveling in—Lyra's dishonesty is that however positively her "artistry" is put, there are clear limits to when Lyra can use it. Specifically, she can use it to "steal the unjust's umbrella"—that is, to counter systemic deception, including and especially that of the Church, "whose assertions about the afterlife are shown to be manipulative and misleading lies."[97] In other words, she can return dishonor for dishonor, and at least partly offset the power advantage held by the stronger, older, more powerful adults who set no limits on their own actions.

Likewise, the constitutionally deceptive Mrs Coulter, who is authorially despised for most of the trilogy and whose tactics, as Woods notes, "ultimately are revealed as a projection of the life-denying, freedom-killing, world-destroying institutional religion responsible for a great deal of suffering in that world and this,"[98] receives sanction to deceive God's regent, Metatron, in order to save Lyra's life. Metatron "outweighs" Mrs Coulter by about the same order of magnitude that Mrs Coulter does Lyra: "Facing them was a being made of light…was he clothed? Did he have wings? She couldn't tell because of the force of his eyes."[99] In this context, when God's most powerful angel exercises no restraint on his divine capabilities, the human Mrs Coulter is implicitly lauded for "lying with her whole life" to trick Metatron, using her history of having "betrayed and intrigued and gloried in [her] treachery to convince the prince of angels that she has no shred of protective love for her daughter."[100]

In keeping with the parameters on moral immorality, Lyra is expected or forced to tell the truth when she is dealing with protagonists who are part of her "team"—for instance, in dealing with her gyptian foster-family, or when she is telling the ghosts in the land of the dead about the beauty of the living world above them. Moreover, from this point forward in the governance of the land of the dead, the ghosts themselves are required to honor the reality of their bodily, sensory experiences as living people in order for the harpies to consent to lead them to the window in the world where they can exit and dissolve to be rejoined atomistically with their daemons. They are told, "You must tell [the harpies] stories. They need the truth. You must tell them true stories, and everything will be well."[101]

Most importantly, with the Authority overcome and Lord Asriel and Mrs Coulter eternally falling into an abyss with Metatron, Will and Lyra encounter a serial test of honor. This comes in the form of several sudden and drastic discoveries that test their ability to make self-limiting sacrifices as leaders of the new order of things. First they learn that, in what seems clearly to be an analogy of eco-niche principles and locality on Pullman's part, their daemons will sicken and die if kept away for more than a few years from the world in which they were born. The living beings of a world must, as Will's father has told them, "build the Republic of Heaven where [they] are, because…there is no elsewhere."[102]

When they first learn of this limitation, the children plan to skirt it by cutting a window between their two worlds and toggling between the two, keeping both daemons healthy. However, they then learn that each window creates a soul-eating Specter and lets out the life-enhancing Dust, and so the "subtle knife" that Will has so freely used in service of overturning both the Authority and Lord Asriel's equally hubristic ambitions abruptly becomes an agent of ill.

The ineluctable conclusion is that they must refrain from cutting any more windows, close all that they know of, and break the subtle knife so that no more harm can be done. At first they resist this sudden demand on their theretofore untapped sense of honor. Will urges Lyra, "Let's try and remember [the explanation] exactly. There might be a way through. There might be a loophole."[103] When their daemons impress upon them that *all* of the windows must be closed, Will tries another dodge, suggesting that they keep just one open, for the two of them. However, this distresses the daemons immensely. At that point, Will and Lyra each offer to come live a drastically shortened life in the other's world. And they face one more great temptation when the angel Xaphania tells them that if Will and Lyra help create more Dust, they can replace any that is lost through a window. This would allow them to leave one window open. Will is excited, but Lyra, the "second Eve," has the strength to reject this new temptation for the greater good of humanity: she sees that they "must leave it open for [the dead]!"[104]

Thus, the new system, in several worlds, begins with the principle of self-limiting honor at the forefront. In place of the arrogant overreach of the Church or Lord Asriel's equally destructive ambitions for rationalist secularity, there are three figureheads—Lyra, Will, and Mary Malone—whose "agency derives not from…limitless freedom, but from love and responsibility" and who understand "ways in which limits can work to restore damaged ecosystems."[105]

The Other Wind begins with and revolves around a scenario of tit-for-tat dishonor. The dragons, who have long held by their oath to stay in the western reaches and leave human habitation unmolested, have now broken it. Or rather—and significantly—the *youngest* generation of dragons has decided not to continue to return honor for dishonor as countless generations of dragons have before them. The "young dragons ...say men are oath breakers who stole the dragons' lands. They tell one another that... they will wait no longer, but will drive men out of all the western lands."[106] They have begun to come during the harvests and set the fields on fire, killing the farmers in the process. This is indeed a show of the previously honorable deciding to act on their power advantage. As King Lebannen points out to his courtiers, "Against the anger of these creatures, if indeed they were to come against us in any number, we have no true defense... Therefore we must try to find out the cause of their anger and, by removing it, make peace with them."[107]

The dragons in these three books serve as symbols in multiple directions. While far from pure in morality themselves, and admittedly "jealous and irate,"[108] they represent the pre-Judaic, pre-industrial, perhaps even pre-agricultural aspect/generations of humans—the part that is not possessive and that has the innate immortality of understanding their place in the scheme of life, death, and existence. They know but do not exploit the language that identifies all entities in creation by their true names and their true essences. They also represent the ways of knowing and being of non-human nature, with their different comprehension of what it means to be alive.

The young dragons' anger, the group eventually learns, is directed toward the western rune-makers—"oath-breakers ...thieves, traitors!" who "stole half [the dragons'] realm from [them], walled it away from life and light, so that [they] could live there forever."[109] As the dragon-woman Orm Irian spits at the assembly of mages and commoners on Roke near the end of the novel, "You own the earth, you own the sea...You wanted land to own. You wanted things to make and keep. And you have that. That was the division, the *verw nadan*. But you were not content with your share. You wanted not only your cares, but our freedom."[110] With the mage Cob's breach between the lands of the living and the dead, as told in *The Farthest Shore*, the dragons have begun to remember this ancient wrong of humans breaking their oath to forgo immortality. They remember also that the western humans' dim waste land of dead souls was once part of their own realm, a previously beautiful land where the dragons could drink from the fires of the volcanoes and fly on "the other wind," out of time and body.

Irian and her dragon kin will be only temporarily placated by being reminded that the current generation of mages, royalty, and common folk are not the ones who stole the land and wind from the dragons, and that they too are paying the price for the broken oaths. The mages, with their "wizardries of death" and "spells of immortality,"[111] must expunge immortalism from their repertoire of power, and dismantle the land of the unliving created by it, or lose all to the scouring of the dragons.

This means the dismantling of the whole of the Western power structure, the entire western paradigm—for the powerful mages who succeeded the rune-makers have, wittingly or unwittingly, instilled this dishonorable mentality deep into the belief systems and social structures of society. The current Master Summoner of Roke, whose particular power is the ability to cross into the land of the dead and to call dead souls out of it, epitomizes the intransigence of dishonor even among those who have spent their lives immersing themselves in the crucial necessity of self-limitation. Seeing the village wizard Alder dying before him, the Summoner brings him back to life, even though Alder's mission is to journey to the land of the dead to help tear down the wall. His subsequent exchange with King Lebannen, mild in tone and brief as it is, summarizes the change in attitude that these events have necessitated and ushered in:

> "Why did you summon him back?" the king asked...
> After a long time the Summoner said, grimly, "Because I had the power to... I was wrong. But it is not right to want to die...For the very old, the very ill, it may be. But life is given us. Surely it's wrong not to hold and treasure that great gift!"
> "Death also is given us," said the king.[112]

Like Lyra's deceitfulness and instinctive understanding of the alethiometer, the young dragons' reciprocal lack of self-limitation—which is not entire, as they will not kill a person unless that person tries to kill them—is given authorial license only on a contingent basis. They may use it to counter the equal and dishonorably used power of the Rune-makers and most power-hungry mages, and to dismantle the male-supremacist systems that prevent women's knowledge and women's power from aiding in the governance of the land. When the dead have been released from the dry land, the Kargish princess released from her veils and subservience to rule as an equal with King Lebannen, the mages of Roke forced to acknowledge the powers of women and wizardly inferiors, and humans revalued as "animals that speak"[113]—when this wholesale remaking has occurred at the end of *The Other Wind*, the dishonorable scouring dragons go as well. "They're all gone...as that shadow place and all the shadows in it rejoined the world of light, so they regained their true realm."[114]

Thus, like Pullman, Le Guin appears to say that the honor that we have tried to inculcate in the next generations is the very thing allowing the nature-severing and dishonorable power systems to stay in place. For that sense of honor, self-restraint, and species-humility to take effect in the Western human populations, with the total rejection of attempts at immortality as its proof, we must untie our hands for a limited time and let ourselves respond to the powerful measure for measure. That done, species honor will be built into the new dispensation, and will be effective.

The theme of the powerless coming into their power and exercising it over the dishonorable crops up in all three of the newer Earthsea books. At the end of *Tehanu*, when, powerless, Ged and Tenar are ensorcelled by the immortalist wizard Aspen, Ged forced to voice his powerlessness over and over for Aspen's delight; Tenar kicked, made to crawl and eat dirt, and called a "bitch" repeatedly; and finally the pair of them taken to a cliff, with Aspen intending to make Ged push Tenar to her death, Tehanu remains idle no more. She uses her kinship to the dragons to summon the great elder dragon Kalessin. Kalessin incinerates Aspen and the aristocratic patriarch of the Isle of Gont, who has been kept unnaturally "alive" through immortalist wizardry.

Similarly, in the long short story "Dragonfly," the girl Irian, subjected to her embittered father's drunken grievances along with his slobbering caresses, tries to enter the male-exclusive community on Roke Island, but five of the nine great mages shun her and see her female presence as challenging their very way of life and Roke's existence. The Master Summoner Thorion, unnaturally returned in living form from the land of the dead, uses his mage-power to pruriently watch Irian bathe naked; exposed as malevolent and dishonorable, he challenges Irian to what is essentially a duel. But when he tries to exercise his power over her by summoning her by her name in an attempt to "bind [her] to obey," Irian "tower[s] above him impossibly, fire breaking forth between them, a flare of red flame in the dusk air, a gleam of red-gold scales, of vast wings—then that was gone, and there was nothing there but the woman standing on the hill path and the tall man bowing down before her, bowing slowly down to earth, and lying on it."[115] Thorion is forced into his identity as a dead person, reduced to "only a huddle of clothes and dry bones and a broken staff,"[116] and Orm Irian takes dragon form and ascends to the skies.

With these definitive assertions of power over the powerful in the "second trilogy," Le Guin suggests that women must join with children, the dragons, others among the disenfranchised, and the few men who will willingly abjure their own power, in order to institute a new system free of the hierarchies in which "power…[is] all these arrangements—one above the other—kings and masters and mages and owners."[117] It is this systematized dishonor, from daily interactions to the tunnel-visioned attitudes of even the greatest mages, that has led to there being "something wrong with the world." In unleashing her dragons, Le Guin also takes on western humanity's relationship with nature. By trying to display their mastery over the laws of life and death, despite all the legitimate powers they have, the mages of Earthsea have acted much like both the scientist-technologists and the founders of Christianity, who expanded on the special status of humans found in Judaism (the only major religion with no concept of human immortality) and added to it an increasingly detailed promise of Heaven, Hell, and purgatory. This is the underlying cause of the great disruption that remains in Earthsea even after Ged and Lebannen seal the breach between the living and the dead in the events of *The Farthest Shore*.

No child put forward

Another way in which *The Amber Spyglass* and *The Other Wind* "talk back" to the tradition of ecocentric children's literature is by not exculpating adults or letting them shunt responsibility onto the younger generations. As Bob Henderson et al. say of Dr Seuss's environmental tract *The Lorax*, many books of the 20th century demonstrate the "standard adult practice" of "wrongly burden[ing] youth while letting adults off the hook," thereby "abdicating responsibility to the younger generation" and leaving crises "for the next generation to resolve, with no guidance from the generation that used the resources."[118] These novels of the 21st century, an era in which environmental disaster has gone from local to global to biospheric, can be seen as the younger generation holding the older generation to account.

The children in these books do indeed, like their real-life readers, inherit the environmental, psychological, and existential damage of the Western belief systems and power structures—the inheritance both of what removing ourselves from nature has done to it and what it has done to us. However, they show the adults that if they want a solution, it will be one that deposes the adults' paradigms and undoes the system of privileges that those structures confer. For this incomplete solution, which cannot undo the damage but may slow it, these books indicate that the adults must pay at least as much as the children.

Early in Pullman's trilogy, the Master of Jordan College, Lyra's only sincere adult protector among the male-dominated halls of Oxford, says, "[It is] the duty of the old…to be anxious on behalf of the young. And the duty of the young is to scorn the anxiety of the old."[119] However, it is for the most part the irresponsibility, power-mongering, and short-sightedness of "the old" that Lyra scorns. She calls her father, the "renegade theologian" Lord Asriel, on his immorality, clearly feeling no filial duty to whitewash his character or align with him: "She felt wrenched apart with unhappiness. And with anger, too; she could have killed her father; if she could have torn out his heart, she would have done so there and then, for what he'd done to Roger. And to her: tricking her: how dare he?"[120]

The children of Pullman's trilogy are no sacrificial lambs. Unbound (until the final pages) by the conventional virtues, they meet harm and danger with retribution, or at least thoughts of it. Lyra gives as good as she gets: referred to by her father as "an insolent child," she retorts, "You en't human, Lord Asriel. You en't my *father*. My *father* wouldn't treat me like that. Fathers are supposed to love their daughters, en't they? You don't love me, and I don't love you, and that's a fact."[121] Convention dispensed with, Lyra interrogates her father for a full explanation of Dust and the Church's attitude toward it. She gives no quarter, and feels no apparent guilt at hitting back with reciprocal callousness.

She feels equally detached from her mother, Mrs Coulter, watching without emotion as she parts from Lord Asriel after he has opened a bridge to the new world in the sky. And when the smooth-talking Sire Charles Lathom steals her alethiometer, she storms at him, "[Y]ou're worse than my mother…You ought to *die*! If I can, I'll make someone kill you. You're not worth leaving alive!" She then spits in his face.[122] Will, hearing that Mrs Coulter has killed Lathom, says with satisfaction, "Well, that's something good, at least."[123]

Pullman overtly acknowledges at a number of points the unfairness of the earlier generations letting environmental and societal catastrophe devolve on the newer generations. Earlier canonical children's literature tends to be silent on this, but Pullman draws attention to it. For example, the former guardian of the subtle knife that can cut through any material and cut holes between the worlds tells Will, "You [the inheritor of the knife] should not be a child. But our world is crumbling, and the mark of the bearer is unmistakable."[124]

Lyra does not venerate the elders who have led the world(s) into these catastrophic conditions. As much to the point, after seeing through the folly and treachery of her father's ways, she works against him as actively as she does the Church-promoting Mrs Coulter. Although she plays no direct part in ensuring that Lord Asriel's new-made Republic of Heaven will fail, she and Will are the ones who decide to break the "subtle knife" that could allow the continued cutting open of windows to allow travel between worlds and thus keep the population of the new Republic from dying of daemon-relocation sickness after just a few years. Lyra and Will sacrifice their new and passionate love for each other to help restore the worlds' ecosystems, but the keepers of power arguably pay a higher price, as the trilogy ends with both the Church and the secularists overthrown, their sources of power diminished or destroyed. Asriel and Mrs Coulter, representatives of these two major adult factions, end up caught in a fate worse than death, plummeting endlessly into the abyss—an authorial indictment, surely, of both characters, as well as a chance for them to redeem themselves by ultimately allying themselves with Lyra's cause.

Pullman challenges the conventional innocence and passive receptivity of children throughout his series. Lyra has her own street wars with gangs in Oxford, which do some real physical and emotional damage. This fact, in part, makes her reconsider the assumptions about Bolvangar, the site of child-daemon "intercision," that "whatever grownups did, however bad it was, kids were different. They wouldn't do cruel things like that." Speaking from her own experience, Lyra notes, "I en't sure now."[125] Likewise, Will, having watched his schoolmates torment his mentally ill mother, responds, "[A]fter that I never trusted children any more than grownups. They're just as keen to do bad things."[126]

This erasure of the moral line between conventionally "innocent" children and corrupted adults gives further license to a "next generation" that is set on making elders pay for their mishandling of the world, although they themselves are (bitterly) accepting that they must be the ones to rectify what they can, and live in, or at least survive, the rest. Will puts the blame squarely on adults for climate change: "They say that people have been interfering with the atmosphere by putting chemicals in it, and the weather's going out of control."[127] Lyra replies matter-of-factly, "Yeah, well, they have...and it is. And we're here in the middle of it."[128]

Keeping in mind the perennial admonishment that children and teens are rarely the writers of the books they read, we can at least read the vindictiveness in *The Amber Spyglass* and *The Other Wind* (among many other dystopian/catastrophe novels for young readers) as authors being allowed, perhaps even encouraged, to be angry on behalf of the younger generations. And surely it is reasonable to infer that their own anger, as former members of "the younger generation," is coming out. This leads them by one narrative sequence or another to kill quite a few of the dishonorable adults in their novels and to have their child characters display a ruthless attitude toward tearing down the power structures.

Earthsea's King Lebannen, only 20 when he follows Ged into the land of the dead in *The Farthest Shore* and only 30 when he learns that "a great change is upon us"[129] at the start of *The Other Wind*, acknowledges the dire state of the world and the need for comprehensive change as directly as Lyra, if more moderately, saying, "We must join together to learn what the change is, its causes, its course, and how we may hope to turn it from conflict and ruin to harmony and peace."[130] The Master Namer of Roke tells a council, "We're out of balance... When and where did we begin to go too far? What have we forgotten, turned our back on, overlooked?"[131] The Kargish Master Patterner thinks, "All the years the patterns of the shadows and the branches and the roots, all the silent language of his forest, had spoken of destruction, of transgression, of all things changed."[132] The young woman-dragon Irian, challenged and threatened by five of Roke's nine great mages, defies them, refusing to give herself over to them and responding to their jibes about "a woman's place" by saying, "My place… My place is on the hill. Where things are what they are."[133] After Irian takes dragon form, the Master Doorkeeper, who gave her entry even though the laws of Roke forbade women to go there to learn, signals the deep and permanent change that Irian has created when he says, "I think we should go to our house, and open its doors."[134]

Council of the wise

Hand in hand with these books' departure from conventions of childhood innocence is their effacement of the often-stark Romanticism-descended division between the mindset and values of children and those of adults as portrayed in children's literature. This erasure retroactively acknowledges today's adults as yesterday's children, and so on back through to Western society's initial eras of concern over the mutually (though arguably asymmetrically) destructive effect of Western society's antagonistic relationship with nature. In books such as *The Secret Garden*, *Tuck Everlasting*, *The Search for Delicious*, *Drowned Ammet*, and even to a degree the original Earthsea trilogy, the child or teen protagonists have almost exclusively held the natural, animal-sympathetic, species-identifying worldview, while adults have been at best indifferent and at worst openly inimical to such outlooks.

True, it is the Tuck parents who initiate efforts to dissuade Winnie from drinking the *elixir vitae*, and in *The Secret Garden* Susan Sowerby presents the possibility of an adult retaining a "childlike" reverence for and engagement with natural surroundings, mostly via her pre-Industrial rural lifestyle. For the most part, however, these earlier books not only deposit the legacy of increasing nature-alienation and exploitation onto "the next generation," but also attribute mostly to youth the sensibility that will make a paradigm shift a priority. This leaves as an open question what happens to that sensibility as the youth becomes an adult and obliged, as we have been since the ascendancy of industrial capitalism and urbanization, either to forsake the bulk of human society in acts of what Lawrence Buell terms "an aesthetic of relinquishment" and "voluntary simplicity"[135] or participate in a lifestyle and system that betrays our earlier values.

In other words, books that present biophilic urges and the apparent absence of them as corresponding fairly strictly to the childhood and adult phases of life perpetuate two problems. Firstly, they suggest that the "affinity for life and lifelike processes" is a matter of individual choice, or at most associated with a developmental phase, rather than an evolutionarily predetermined biological imperative whose thwarting comes at a high psychological cost. Secondly, they facilitate a silence on the internal disjuncture that arises when this biological imperative is put into conflict with the imperative of human sociability. This perpetuates both the tacit assumption of generational amnesia and the societal expectation that we will keep silent about our dismay over self-apostasy—in other words, we will keep our anomie to ourselves. For this reason, there has long been a phenomenon of resetting at zero, each generation, even while the socio-psychological and environmental damage has increased dramatically.

In *The Amber Spyglass* and *The Other Wind*, there are, to be sure, adults who are depicted as caring nothing about the degradation of the world and humanity's increasing divorce from its evolutionary support system. These are not only villains such as Lord Asriel, the Master Summoner Thorion, and the child-rapist Handy: sympathetic characters such as the witch Serafina Pekkala and King Lebannen are also shown as focusing their concerns more on disturbances to society than on human-nature relationships or the state of the environment.

However, there are in each set of texts a number of adults who show deep concern for environmental, human-natural, and eco-spiritual issues. Such characters show that these priorities and foci can and do survive into adulthood, and that they can be expressed as a central part of society reconfiguration.

The nun-turned-physicist Mary Malone, for instance, receives extensive space in the latter parts of Pullman's trilogy to convey the central battle over Dust in ecological terms. Throughout the final book, the theme of "human" (represented by the pacifist sentient species the *mulefa*) interdependence with nature forms an obbligato with Pullman's moral revision. The destruction of the trees in the *mulefa*'s world and of the *mulefa* themselves, along with the melting of the polar caps in Iorek Byrnison's world (Lyra's original one) and the climatological disturbances in Will's world (our own), are the trilogy's most overt commentary on the catastrophic results of Judeo-Christian culture's self-arrogation and the destructive development of industrial capitalism which, in many ecocritics' view, that culture has sanctioned. The fact that it is an adult, rather than a child, who experiences this tragic destruction allows Pullman to equate the interaction between Dust and the trees with the sexual pleasure that constitutes part of "original sin." He thus depicts the sexual nature of much of life that religious prudishness has sanitized away: "There was a light breeze, which lifted a faint scent out of the flowers and rustled the stiff leaves, and Mary imagined a huge, dim benevolence holding her up, like a pair of giant hands. As she lay in the fork of the great branches, she felt a kind of bliss she had only felt once before; and that was not when she made her vows as a nun."[136]

Just as Mary is having an orgasmic experience lying in the trees, the trees are likewise having sex with the Dust, or, as the *mulefa* call it, the *sraf*. As Mary explains it, "[T]he trees are filtering the *sraf* as it moves through them, and some of it is attracted to the flowers. I could see it happening: the flowers are turned upwards and if the *sraf* were falling straight down, it would enter their petals and fertilize them like pollen from the stars."[137]

Mary's perspective gives some indication of how the impressionistic environmental concerns of childhood, such as Lyra's and Will's, could develop into more informed and mature analyses. These give the implications more urgency than those of childhood, not less. Mary's ability, through scientific training and sexual experience, to identify the innate need, goodness, and rightness of sex for all sexually reproducing strains of nature, including humans, performs the trilogy's main equating of humans with other nature, working against Pullman's otherwise humanistic and anthropocentric view. It offers, as well, a sense of the lifestyle and career choices necessary to live as an adult at least partly in accord with the animalistic, biophilic priorities permitted in childhood. Mary has rejected life as a nun to be able to participate in life as a sexual, sensual, body-inhabiting being, contrary to many (though not all) strains of Christian doctrine. Mary's arc may teach the child reader that one way of staying true to the nature-focused outlook of childhood is to remain inside the body, self-possessed, and honor sensory and sexual urges.

This complements the privileging of sensory, animalistic living that Pullman creates when Lyra and her companions strike a new deal with the harpies in the land of the dead: that the dead souls can go to the window in the afterlife, and dissolve to rejoin their daemons and the rest of the living universe, only if they "tell the truth about what they've seen and touched and heard and love and known in the world."[138]

The alliance between Mary Malone, Lyra, Will, and other enlightened adults in the trilogy—the bear Iorek Byrnison, the shaman John Parry, the explorer Lee Scoresby, the angel Xaphania—is in some respects a typical trope for fantasy, a gathering of companions for the young main protagonist. We see this in books from the Narnian and Prydain chronicles to The Lord of the Rings, Harry Potter, and the Guardians of Ga'hoole. The difference is that instead of enlisting the young to help defend the status quo against antagonists determined to overthrow it, Pullman's adult companions join the children in their efforts to overthrow a status quo that has consistently and increasingly led away from first principles of authentic (i.e., evolutionarily prescribed) living and tipped the world into chaos.

Although, as I have said earlier, Pullman's trilogy does not primarily concern itself with ecological crisis or human-nature crisis, this crisis does form the backdrop and act as an intensifier—sometimes mimetic, sometimes symbolic—of the atmosphere of awryness and impending catastrophe in the books. In the end, the new status quo is not only about moral permission to experience and enjoy the sensory and sexual realities of the world, but to be rejoined—perceptually and literally—with "the rest of the living universe," repositioned as on par with the other animals and within the living matrix rather than artificially removed by illusions of immortality. While it is Lyra who goes to the land of the dead and persuades the ghosts to follow her along the edge of the abyss, and it is Will who cuts the window, enlightened adults, working toward the same end, are indispensable to the revolution's success. These include the mature Gallevespian Tialys, who strikes the bargain for the new process with the harpies; Mary Malone, who clearly conveys that the fall of the Church means humans' re-entry into the cycle of life; and the angel Xaphania, who reintroduces to the new dispensation the need for systematized self-restraint and strict limitations.

In this assemblage of ages, races, species and states of being, children are treated with parity, respected for their particular gifts or roles rather than because they are children. Adults, similarly, are portrayed as villains, heroes, or combinations of the two according to their traits and values, not their developmental states. Lord Asriel receives help and admiration from sympathetic characters, including Will's shaman father John Parry and Lyra herself, when he is acting to topple the Church. He receives censure and becomes an antagonist when he severs Lyra's friend Roger from his daemon, when he shows his motivation to be as much vainglory as a desire to liberate society from the tyranny of the Church, and when he disregards the environmental disasters that his actions unleash in multiple worlds. In short, many conventional ways of portraying children and adults according to their age groups are dispensed with, and Pullman instead categorizes—often indeterminately—by ideology and character.

This rubric shows even more strikingly in Le Guin's latter Earthsea books. They make little distinction between characters' ages, and have as antagonists only representatives of the traits that are facilitating the ruination of Earthsea: the immortalists Alder and Thorion; the rapist Handy; the five male-supremacist mages who range themselves at Irian. One of the trilogy's central protagonists, Tehanu, as a child acts infrequently and reluctantly to fulfill her prophesied role as an agent and vehicle of change. As a young woman giving counsel to the king, she discovers without pride or clear foreknowledge that she can outdo even the greatest of the mages and kings in mediating with dragons, as she can tap into her dragon-self to speak in the Language of the Making and claim as her father the eldest dragon, Kalessin. Hence, there is little or no correspondence between age and innocence or experience, nor between childhood and connection with nature or a different mode of being human.

Similarly, Tehanu's foster-mother, Tenar, as a teenager repudiates her identity as "the Eaten One,"[139] vehicle-priestess of the Kargish God-King, and uses her status to overcome the Old Power of the Earth where Ged is helpless against them even at the height of his powers. However, when Le

Guin picks up her story again to explore "what the 'dry land' was, why it existed, who the dragons are and what is their relationship to human beings,"[140] Tenar has only modulated, not rejected, her understanding of those powers. As a Karg, she has retained a knowledge of the earth's Old Powers than the western residents of the Hardic culture who produce mages and wizards and therefore have broken the covenant with the dragons. While her experience frees her from the kind of superstitions ruling the Kargish princess Seserakh about "the accursed-sorcerers,"[141] Tenar does retain the knowledge that people and animals had originally been the same order of being until one set of humans began to achieve wealth, power, and possessiveness. As an older woman, with white hair and an admiring envy of the youth on parade, she comes back into her Kargish heritage, telling King Lebannen, "[M]y people don't go where your people do when they die. That dry land...it's not where we go. The princess, and I, and the dragons."[142]

Thus, through all the phases of her life, Tenar deepens her analysis of humans' relationship with nature and their self-identification in that relationship. Like Mary Malone, she has the sexual experience that other main characters lack (Ged and Lebannen, for instance, have long been virgins, as are all of the Roke mages), and it is partly this that enables her to deconstruct the need for power, the systemic male dishonor, and both men's and women's fear of women's power. "Why are men afraid of women?" she asks in *Tehanu*. "Women seem to fear their own strength, to be afraid of themselves...If power were trust...if it weren't all these arrangements—one above the other—kings and masters and mages and owners—It all seems so unnecessary. Real power, real freedom, would lie in trust, not force."[143]

Tenar and Tehanu from the first interact as equals despite their ages and nominal powers. King Lebannen re-pairs with the much older Ged, spiritually if not in body. These books thus present models of adulthood in which childhood allegiances to a more balanced set of values are strengthened and deepened by the conscious understanding, sexual experience, life experience, and social positions that confer insight. The wise adults in both trilogies see the need for an overthrow of the systems that are keeping western society at odds with natural systems.

Common cause

As I remark in my opening to this chapter, one new commonality and newly central focus in Pullman's and Le Guin's texts is the markedly feminist, and even ecofeminist, slant. Greg Garrard defines ecofeminism as "involv[ing] the recognition that [anthropocentrism and androcentrism] share a common 'logic of domination'...or underlying 'master model', that ...should suggest common cause between feminists and ecologists."[144]

Karen J. Warren modifies this with the observation that "Just as there is not one feminism, there is no one ecofeminism."[145] Both Pullman and Le Guin display some degree of belief in biological gender determinism that places them outside of ecofeminism, just as their privileging of the human species has them somewhat at odds with the ecological movements in general. Both authors in fact to some extent assign a biological destiny that many feminists reject. Pullman displays this when he has Mrs Coulter act consistently against the interests of her stated aims in order to protect her daughter, whom she gave up at birth and barely knows, despite being, as the angel Metatron puts it, characterized by "corruption and envy and lust for power. Cruelty and coldness... .Pure, poisonous, toxic malice"—in short, being "a cesspit of moral filth."[146] Her maternal urge, in distinction from Lord Asriel's continued and absolute lack of paternal instinct, overrides all these components of her character: she says to Lord Asriel, "I wanted [Metatron] to find no good in me, and he didn't. There is none. But I love Lyra...[N]ow I love her so much my heart is bursting with it."[147]

Le Guin, meanwhile, depicts most men, though not all (the village sorcerer Alder and the Kargish Master Patterner Azver being notable exceptions), as gaining power through force rather than trust. Ged avers, "All the greatness of men is founded on shame, made out of it."[148] Like Pullman, she assigns some amount of biological destiny to women, having Tenar think,

> How men feared women! Not as individuals, but women when they talked together, worked together, spoke up for one another—then men saw plots, cabals, constraints traps being laid. Of course they were right. Women were likely, as women, to take the next generation's part, not this one's; they wove the links men saw as chains, the bonds men saw as bondage.[149]

Despite all this, both authors present a remarkably nuanced ecofeminism. Taken as a whole, it is one that avoids the common essentialist linking of women and nature that offers "a mirror-image of patriarchal constructions of femininity that is just as limited and limiting"[150] and instead finds more or less the "third position" of ecofeminism that Val Plumwood advocates for, one that exposes and mitigates the often-buried but still pernicious "dualistic dynamic" operating in both the deep ecology and social ecology movements.[151] Neither author presents women in themselves as either more or less intimately connected with nature than men. Rather, women, like children (and children's literature), are in positions to subvert power structures because they are marginalized, scorned, and dismissed. Le Guin herself articulates this when she says of Tenar, "Her insignificance is her wildness. What she is and does is 'beneath notice'—invisible to the men who own and control, the men in power."[152]

All of these marginalized and overlooked characters use the leverage they get from being underestimated to carry out their insurrections. Lord Asriel, for example, has ample opportunity to neutralize Lyra at many points, and to take the all-important alethiometer from her, but he casually lets her and her golden compass leave, contemptuously saying that "the fuss she's caused is out of all proportion to her merits. An ordinary English girl, not very clever...I can't waste any more time or resources on her."[153] Mrs Coulter, too, is made careless with Lyra by the girl's undistinguished status as a female, an urchin, and an uneducated person. Only after Mrs Coulter tortures a witch and learns of their prophecy that Lyra will be a second Eve, one whose temptation will determine the history of the universe, does she start chasing Lyra and trying to keep her from that destiny.

Earthsea's Tenar, Tehanu, Irian, and Seserakh, similarly, are assigned no sexually determined affinity for nature or special understanding of how humanity can realign and change the sickness of the world. Their special relationship comes from other aspects of their identity—and insofar as it stems from their gender, it is because that relationship has essentially been forced on them by the male-supremacist power structure. "No woman can become Archmage," Ged reminds Tenar. "Both manhood and magery are built on one rock: power belongs to men."[154]

The women of this "second trilogy" have not been allowed to participate fully in the violation of the *verdunan*, any more than they have been allowed to share the power men got from their side of the bargain. When the Kargish Master Patterner of Roke issues the oblique prophecy that the mages will find their answer from "a woman of Gont,"[155] they assume that this means the woman will lead them to the new Archmage, and they are scornfully offended at the interpretation that the woman might be important in herself, or perhaps even be the Archmage. Only Ged, newly wise in the ways of the powerless, has the humility to point out, "Well, the Patterner didn't say that a woman on Gont was to be Archmage. [He just said] 'A woman on Gont."[156] So it is that Tehanu, shunned and abhorred, is able to call the dragon Kalessin to incinerate the immortalist Aspen—or "the one called Aspen, whose [true] name was Erisen, and whom she saw as a forked and writhing darkness."[157] No one suspects that she might have powers beyond those of the greatest mage, with the Language of the Making as her mother tongue, the ability to parley with dragons, and her innate knowledge of everyone and everything's "true name" giving her unlimited power if she chooses to wield it.

Tehanu's dragon-woman sister Irian, daughter of a dispossessed former aristocrat, is spurned by five of the nine mages of Roke, and taken lightly by the contemptuous immortalist Master Summoner Thorion. Hence, despite warnings from the Master Patterner and the inherent attribution of power to Irian that the Master Doorkeeper makes by letting her into the male-only stronghold of Roke, Thorion challenges Irian and lets her lead him to Roke Knoll, where all things reveal what they truly are. There Thorion takes his true nature as a dead man, while Irian takes hers as both dragon and woman.

Meanwhile, Seserakh, though daughter of the self-appointed High King of the Kargs, who has deposed the God-King and ended an ages-long theocracy, is seen and treated as "a stupid ignorant desert woman. A *feyagat* [wearer of the burqa-like veil]."[158] Discounted even by the normally equable and honorable King Lebannen, she is overlooked as a source of important knowledge about the relationships between dragons, humans, life, death, and rebirth. Tenar has to relay this vital news—that neither Kargs nor dragons go to the "dry land" after death—to Lebannen before he can swallow his pride, master his fear of being trapped by a woman, and talk to Seserakh as an equal for the first time.

Thus, it is the contemptible status conferred upon women by men, not women's power in itself, that makes women in these books the able saboteurs that they are—the ones who initiate the toppling of the immortalist, anthropocentric, and androcentric status quo. Women might lead the revolution, but not because they are women.

Indeed, perhaps one of the most striking aspects of these two trilogies is the democratic manner in which they assemble their large sets of protagonists. This implies three perceptions on Le Guin and Pullman's parts. The first is that they see potential for a nature-allied revolution in almost every quarter, across class, culture, gender, and status lines. The second is that they see a massive, system-wide effort as necessary if the entrenched Abrahamic-based patriarchy is to be deposed and a humbler, pre-/post-Christian way of living instituted. Pullman comes out and says as much as this through his character King Ogunwe. This African king has joined forces with Lord Asriel and his host of rebel angels because he sees the hope of destroying the tyranny of the Church. As he explains,

> The Authority [God/the Church] is not the creator…[A]t some point the Authority took charge, and since then, angels have rebelled, and human beings have struggled against him too. This is the last rebellion. Never before have humans and angels, and beings from all the worlds, made a common cause. This is the greatest force ever assembled. But it may still not be enough. We shall see.[159]

The third implication is that combined and cooperating perspectives and ontologies are the best way to achieve comprehensive reform. In *The Other Wind*, Tenar laughs at the eclectic group of beings gathered on a ship to journey to the Immanent Grove on Roke. "What a ragbag you are bringing them, to be sure!" she teases King Lebannen. "A sorcerer with nightmares, a wizard from Paln, two dragons, and two Kargs."[160] Later Tenar thinks, "So, here in the center where the earth's powers met, the human powers had also met together: a king, a princess, the masters of wizardry. And the dragons."[161] The use of the word "ragbag" is presumably a reference to Le Guin's own essay "The Carrier Bag of Fiction" (1986), a work of ecofeminism canonical enough to be included in Cheryll Glotfelty and Harold Fromm's American ecocritical bible *The Ecocriticism Reader*. In it, Le Guin argues against a philosophy of writing (and therefore of culture) in which "the Hero has decreed through his mouthpieces the Lawgivers... that the central concern of narrative, including the novel, is conflict [and that] the story isn't any good if he isn't in it,"[162] and in favor of "the natural, proper, fitting shape of [a] novel [being] that of a sack, a bag. A novel is a medicine bundle, holding things in a particular, powerful relation to one another and to us."[163]

No easy answer

In the end, both Le Guin and Pullman resolve the moral, spiritual, and sociological problems of their texts more easily and completely than the ecological and environmental ones. In His Dark Materials, the self-restraint of environmental awareness, born of a heightened consciousness of ecological limitations, resource depletion, and climatic disruption, to some degree substitutes for the behavioral self-restraint demanded by the Church. Sarah Cantrell points out that "[i]n the trilogy, limitlessness is understood as hopelessly corrupt,"[164] with Pullman suggesting through his moral equation of Lord Asriel and Mrs Coulter that socioreligious oppression and ecological destruction extend from the same hubristic actions. Asriel, who believes in "an immediately consumable elsewhere that is his for the taking,"[165] and Mrs Coulter, who is an agent of "a vicious Church with limitless power,"[166] end up in an eternal freefall, with their ambitions similarly stymied.

However, beyond the edict that we must stay in the place where we belong—that is, the world that we and our daemons were born in—Pullman offers no program of change to counter the apocalyptic environmental disasters that he depicts human arrogance as creating. In fact, the Enlightenment-style secularism that his protagonists have achieved through the overthrow of the Church is, historically, the first step toward the modernization, industrialization, and urbanization that constitutes a great deal of the perceived ecological and human-spiritual crisis today. Therefore, despite the trilogy's ultimate censure of Lord Asriel, the same technologically ambitious, essentially humanist outlook seems to be set in place and seems likely to continue from there, leading us back to basically this same place of psychological severance and environmental/atmospheric crisis.

Le Guin, having presented as the original problem only the fact of "imbalance" and "having gone too far,"[167] has a lighter burden to manage. Like Pullman, her books suggest that "what the typical male hero must learn is *limits*."[168] They also suggest that once women, children, the poor and humble, and the agents of wild nature are included in the workings of power, and the immortalist foundations dismantled, and western/westernized society re-instilled with a species self-perception that brings them to the same level as other animals, the principles of self-limitation and the checks and balances that prevent disastrous overreach will naturally follow. With the wisest among the powerful guiding us, as King Lebannen says to the council at Roke, we may be able to foresee the events that stem from that overreach, without being overwhelmed by them.

Thus, neither *The Amber Spyglass* nor *The Other Wind* fully qualifies as what Richard Kerridge calls a "narrative of resignation" in the sense of "setting their calamities beyond the reach of remedy."[169] While it is true that both books eschew the belief that "a pragmatic, incremental environmentalism, allied to other social movements, might begin to turn the tide,"[170] they present programs of a sort to stem, if not reverse, the increasing signs that "something [is] wrong with the world."[171] Those programs are nothing short of a holistic change of the world order, with at their heart the undoing of unnatural afterworlds created by immortalist men as both a symptom and an agent of their need for domination.

However, as assertive and newly explicit as these texts are about the state of the western world and its effects on the globe and atmosphere, they are potentially more discouraging than otherwise, in that they lucidly present the need for thoroughgoing change but offer solutions that are unlikely to be achievable in real life.

Notes

1 E. O. Wilson, *Biophilia* (Harvard University Press, 1984), 1.
2 Wilson, *Biophilia*, 80.
3 Wilson, *Biophilia*, 81.
4 Al Gore, speech to New York University School of Law. September 18, 2006. NYU Law School. Qtd in Kerridge, "Narratives of Resignation," 87.
5 Le Guin, *The Other Wind* (Ace, 2003), 142.
6 Le Guin, "Dragonfly," 247.
7 Arthur Bradley and Andrew Tate, *The New Atheist Novel: Philosophy, Fiction and Polemic after 9/11* (Continuum, 2010), 62.
8 Pullman, *The Amber Spyglass*, 457.
9 Pullman, *The Subtle Knife*, 526.
10 Philip Pullman, *The Golden Compass*, ed. His Dark Materials Omnibus (Knopf, 2007), 277.
11 Pullman, *Amber Spyglass*, 287.

12 Sarah K. Cantrell, "Letting Specters in: Environmental Catastrophe and the Limits of Space in Philip Pullman's *His Dark Materials*," *Children's Literature Association Quarterly* 39, no. 2 (2014): 248.
13 Kay Sambell, "Carnivalizing the Future: A New Approach to Theorizing Childhood and Adulthood in Science Fiction for Young Readers," *The Lion and the Unicorn* 28, no. 2 (2004): 247–48.
14 Richard Gooding, "*Clockwork*: Philip Pullman's Posthuman Fairy Tale," *Children's Literature in Education* 42 (2011): 310.
15 Elaine Ostry, "'Is He Still Human? Are You?': Young Adult Science Fiction in the Posthuman Age," *The Lion and the Unicorn* 28, no. 2 (2004): 240.
16 Gooding, "Clockwork," 320.
17 Helena Feder, *Ecocriticism and the Idea of Culture* (Routledge, 2016), 226.
18 Feder, *Ecocriticism*, 227.
19 Sambell, "Carnivalizing the Future," 253.
20 Sarah K. Cantrell, "'Nothing Like Pretend': Difference, Disorder, and Dystopia in the Multiple World Spaces of Philip Pullman's *His Dark Materials*," *Children's Literature in Education* 41 (2010): 314.
21 Karen J. Warren, "The Power and the Promise of Ecological Feminism," *Environmental Ethics* 12, no. 2 (1990): 123.
22 Val Plumwood, *Feminism and the Mastery of Nature* (Routledge, 1993), 2.
23 Lisa Hammond Rashley, "Revisioning Gender: Inventing Women in Ursula K. Le Guin's Nonfiction," *Biography: An Interdisciplinary Quarterly* 30, no. 1 (2007): 27.
24 Pullman, *Golden Compass*, 290.
25 Lynn White, Jr., "The Historical Roots of Our Ecologic Crisis," in *The Ecocriticism Reader*, eds. Cheryll Glotfelty and Harold Fromm (University of Georgia Press, 1996), 10.
26 White, Jr., "The Historical Roots of Our Ecologic Crisis," 12.
27 Gooding, "Clockwork," 309, 318.
28 Graham Holderness, "'The Undiscovered Country': Philip Pullman and the 'Land of the Dead.'" *Literature and Theology* 21, no. 3 (2007): 288.
29 Holderness, "'The Undiscovered Country'," 288.
30 Bradley and Tate, *The New Atheist Novel*, 58.
31 Philip Pullman, *The Amber Spyglass* (Del Rey, 2001), 365.
32 Millicent Lenz, "Introduction." In *His Dark Materials Illuminated: Critical Essays on Philip Pullman's Trilogy*, ed. Millicent Lenz (Wayne State University Press, 2005), 5.
33 Rashley, "Revisioning Gender," 27.
34 Sandra J. Lindow, "Becoming Dragon: The Transcendence of the Damaged Child in the Fiction of Ursula K. Le Guin," *Extrapolation* 44, no. 1 (2003): 32.
35 Rashley, "Revisioning Gender," 28.
36 Le Guin, *Other Wind*, 199.
37 Warren G. Rochelle, "The Emersonian Choice: Connections between Dragons and Humans in Le Guin's Earthsea Cycle," *Extrapolation* 47, no. 3 (Winter 2006): 417.
38 Rochelle, "Emersonian Choice," 417.
39 Cantrell, "Nothing Like Pretend," 235.
40 Le Guin, *Other Wind*, 200.
41 Rochelle, "Emersonian Choice," 423.
42 Le Guin, "Dragonfly," 231.
43 Darko Suvin, "On U. K. Le Guin's 'Second Earthsea Trilogy' and Its Cognitions: A Commentary," *Extrapolation* 47 (2006): 492.

44 Stephanie Guerra, "Colonizing Bodies: Corporate Power and Biotechnology in Young Adult Science Fiction," *Children's Literature in Education* 40 (2009): 282.
45 Le Guin, *Other Wind*, 195.
46 Pullman, *Amber Spyglass*, 485.
47 Holderness, "Undiscovered Country," 279.
48 Holderness, "Undiscovered Country," 280.
49 Suvin, "Le Guin's 'Second Earthsea Trilogy'," 491.
50 Suvin, "Le Guin's 'Second Earthsea Trilogy'," 497.
51 White, Jr., "Historical Roots," 11.
52 David Ehrenfeld, *The Arrogance of Humanism* (Oxford University Press, 1978), 5.
53 Le Guin, *Other Wind*, 193.
54 Le Guin, *Other Wind*, 195.
55 Pullman, *Amber Spyglass*, 286.
56 Plumwood, *Feminism and Mastery*, 23.
57 Plumwood, *Feminism and Mastery*, 196.
58 Pullman, *The Subtle Knife*, 526.
59 Pullman, *Amber Spyglass*, 286.
60 Pullman, *Amber Spyglass*, 278.
61 Ursula Le Guin, "The Carrier Bag Theory of Fiction," in *The Ecocriticism Reader*, eds. Cheryll Glotfelty and Harold Fromm (University of Georgia Press, 1996), 152–53.
62 Ursula Le Guin, *Tehanu* (Simon Pulse, 2001), 248.
63 Le Guin, *Other Wind*, 200.
64 Le Guin, *Other Wind*, 122.
65 Le Guin, *Other Wind*, 193.
66 Le Guin, *Other Wind*, 142.
67 Suvin, "Le Guin's 'Second Earthsea Trilogy'," 497.
68 Le Guin, *Other Wind*, 74.
69 Le Guin, *Other Wind*, 74.
70 Le Guin, *Other Wind*, 75.
71 Pullman, *Amber Spyglass*, 283.
72 Pullman, *Amber Spyglass*, 269.
73 Pullman, *Amber Spyglass*, 273.
74 Pullman, *Amber Spyglass*, 286.
75 Pullman, *Amber Spyglass*, 324.
76 Le Guin, *Other Wind*, 184.
77 Le Guin, "Dragonfly," 236.
78 Le Guin, *Other Wind*, 183.
79 Le Guin, *Other Wind*, 205.
80 Le Guin, *Other Wind*, 198.
81 Rabkin, "Immortality," in *Immortal Engines: Life Extension and Immortality in Science Fiction and Fantasy,* eds. Slusser, George, Gary Westfahl, and Eric S. Rabkin (University of Georgia Press, 1996), xiii–xx.
82 Roderick McGillis, "Introduction: Children's Literature and the *Fin de Siècle*," in *Children's Literature and the Fin de Siècle* (Praeger, 2003), xii.
83 Sambell, "Carnivalizing the Future," 248.
84 Cantrell, "Nothing Like Pretend," 234.
85 Pullman, *Golden Compass*, 275.
86 Pullman, *Golden Compass*, 277.
87 Plumwood, *Feminism and Mastery*, 6.
88 Wilson, *Biophilia*, 81.

89 Pullman, *Amber Spyglass*, 28.
90 Kiera Vaclavik, "The Truth in Lyra's Lies," in *The Golden Compass and Philosophy: God Bites the Dust*, eds. Richard Greene and Rachel Robison (Open Court, 2009), 64.
91 Pullman, *Amber Spyglass*, 261.
92 Pullman, *Amber Spyglass*, 85.
93 Lauren Shohet, "Reading Dark Materials," in *His Dark Materials Illuminated: Critical Essays on Philip Pullman's Trilogy*, ed. Millicent Lenz (Wayne State University Press, 2005), 26.
94 Pullman, *Compass*, 207.
95 Pullman, *Compass*, 208.
96 Vaclavik, "The Truth in Lyra's Lies," 64.
97 Vaclavik, "The Truth in Lyra's Lies," 65.
98 Naomi Wood, "(Em)bracing Icy Mothers," in *Wild Things*, eds. Sidney I. Dobrin and Kenneth B. Kidd (Wayne State University Press, 2004), 210.
99 Pullman, *Amber Spyglass*, 355.
100 Pullman, *Amber Spyglass*, 356.
101 Pullman, *Amber Spyglass*, 386.
102 Pullman, *Amber Spyglass*, 325.
103 Pullman, *Amber Spyglass*, 434.
104 Pullman, *Amber Spyglass*, 441.
105 Cantrell, "Nothing Like Pretend," 234–35.
106 Le Guin, *Other Wind*, 94.
107 Le Guin, *Other Wind*, 118.
108 Le Guin, *Other Wind*, 129.
109 Le Guin, *Other Wind*, 194.
110 Le Guin, *Other Wind*, 194.
111 Le Guin, *Other Wind*, 130.
112 Le Guin, *Other Wind*, 119.
113 Le Guin, *Other Wind*, 196.
114 Le Guin, *Other Wind*, 211.
115 Le Guin, "Dragonfly," 245.
116 Le Guin, "Dragonfly," 246.
117 Le Guin, *Tehanu*, 246–47.
118 Bob Henderson, Merle Kennedy, and Chuck Chamberlin, "A Pedagogical Response to *The Lorax*," in *Wild Things: Children's Culture and Ecocriticism*, eds. Sidney I. Dobrin and Kenneth B. Kidd (Wayne State University Press, 2004), 138–39.
119 Pullman, *Golden Compass*, 24.
120 Pullman, *Golden Compass*, 291.
121 Pullman, *Golden Compass*, 271.
122 Pullman, *Subtle Knife*, 418.
123 Pullman, *Amber Spyglass*, 11.
124 Pullman, *Subtle Knife*, 435.
125 Pullman, *Subtle Knife*, 488.
126 Pullman, *Subtle Knife*, 489.
127 Pullman, *Subtle Knife*, 523.
128 Pullman, *Subtle Knife*, 524.
129 Le Guin, *Other Wind*, 189.
130 Le Guin, *Other Wind*, 189.
131 Le Guin, "Dragonfly," 236.
132 Le Guin, "Dragonfly," 239.

133 Le Guin, "Dragonfly," 243.
134 Le Guin, "Dragonfly," 249.
135 Lawrence Buell, *The Environmental Imagination: Thoreau, Nature Writing, and the Formation of American Culture* (Belknap/Harvard University Press, 1995), 168.
136 Pullman, *Amber Spyglass*, 243–44.
137 Pullman, *Amber Spyglass*, 245.
138 Pullman, *Amber Spyglass*, 284.
139 Le Guin, *The Tombs of Atuan*, 7.
140 Bernardo and Murphy, *Ursula Le Guin: A Critical Companion*, 7.
141 Le Guin, *Other Wind*, 104.
142 Le Guin, *Other Wind*, 113.
143 Le Guin, *Tehanu*, 246–47.
144 Greg Garrard, *Ecocriticism* (Routledge, 2004), 23.
145 Warren, "Power and Promise of Ecological Feminism," ii.
146 Pullman, *Amber Spyglass*, 356.
147 Pullman, *Amber Spyglass*, 362.
148 Le Guin, *Tehanu*, 249.
149 Le Guin, *Other Wind*, 118.
150 Garrard, *Ecocriticism*, 24.
151 Plumwood, *Feminism and Mastery*, 2.
152 Ursula Le Guin, "Earthsea Revisioned" (Children's Literature New England, 1993). Qtd in Rochelle, 422–23.
153 Pullman, *Golden Compass*, 178.
154 Le Guin, *Tehanu*, 245.
155 Le Guin, *Tehanu*, 176.
156 Le Guin, *Tehanu*, 245.
157 Le Guin, *Tehanu*, 272.
158 Le Guin, *Other Wind*, 106–7.
159 Pullman, *Amber Spyglass*, 188.
160 Le Guin, *Other Wind*, 170.
161 Le Guin, *Other Wind*, 187.
162 Le Guin, "Carrier Bag," 152.
163 Le Guin, "Carrier Bag," 153.
164 Cantrell, "Nothing Like Pretend," 236.
165 Pullman, *Amber Spyglass*, 237.
166 Pullman, *Amber Spyglass*, 236.
167 Le Guin, "Dragonfly," 236.
168 Millicent Lenz, "Ursula Le Guin," in *Alternative Worlds in Fantasy Fiction*, eds. Millicent Lenz and Peter Hunt (Continuum, 2001), 43.
169 Kerridge, "Narratives," 99.
170 Kerridge, "Narratives," 99.
171 Pullman, *Amber Spyglass*, 420.

Bibliography

Bernardo, Susan, and Graham J. Murphy, eds. *Ursula K. Le Guin: A Critical Companion*. Critical Companions to Popular Contemporary Writers series. Greenwood Press, 2006.

Bradley, Arthur, and Andrew Tate. *The New Atheist Novel: Philosophy, Fiction and Polemic after 9/11*. Continuum, 2010.

Buell, Laurence, *The Environmental Imagination: Thoreau, Nature Writing, and the Formation of American Culture*. Belknap/Harvard University Press, 1995.
Cantrell, Sarah K. "Letting Specters in: Environmental Catastrophe and the Limits of Space in Philip Pullman's *His Dark Materials*." *Children's Literature Association Quarterly* 39, no. 2 (2014): 234–51.
Cantrell, Sarah K. "'Nothing Like Pretend': Difference, Disorder, and Dystopia in The Multiple World Spaces of Philip Pullman's *His Dark Materials*." *Children's Literature in Education* 41 (2010): 302–22.
Feder, Helena. *Ecocriticism and the Idea of Culture*. Routledge, 2016.
Garrard, Greg. *Ecocriticism*. Routledge, 2004.
Gooding, Richard. "*Clockwork*: Philip Pullman's Posthuman Fairy Tale." *Children's Literature in Education* 42 (2011): 308–24. https://doi.org/10.1007/s10583-011-9135-2.
Guerra, Stephanie. "Colonizing Bodies: Corporate Power and Biotechnology in Young Adult Science Fiction." *Children's Literature in Education* 40 (2009): 275–95. https://doi.org/10.1007/s10583-009-9086-z.
Henderson, Bob, Merle Kennedy, and Chuck Chamberlin. "A Pedagogical Response to *The Lorax*." In *Wild Things: Children's Culture and Ecocriticism*, edited by Sidney I. Dobrin and Kenneth B. Kidd. Wayne State University Press, 2004.
Holderness, Graham. "'The Undiscovered Country': Philip Pullman and the 'Land of the Dead'." *Literature and Theology* 21, no. 3 (September 2007): 276–92. https://doi.org/10.1093/litthe/frm025.
Kerridge, Richard. "Narratives of Resignation: Environmentalism in Recent Fiction." In *The Environmental Tradition in English Literature*, edited by John Parham. Ashgate 2002.
Le Guin, Ursula K. "The Carrier Bag Theory of Fiction." In *The Ecocriticism Reader*, edited by Cheryll Glotfelty and Harold Fromm. University of Georgia Press, 1996.
Le Guin, Ursula K. "Dragonfly." In *Tales from Earthsea*. Ace, 2003.
Le Guin, Ursula K. *The Other Wind*. 2001. New York: Ace 2003.
Le Guin, Ursula K. *Tales from Earthsea*. 2001. New York: Ace 2003.
Le Guin, Ursula K. *Tehanu*. 1990. New York: Simon Pulse, 2001.
Le Guin, Ursula K. *The Tombs of Atuan*. 1971. Bantam, 1983.
Lenz, Millicent. "Introduction." In *His Dark Materials Illuminated: Critical Essays on Philip Pullman's Trilogy*, edited by Millicent Lenz and Carole Scott. Wayne State University Press, 2005.
Lenz, Millicent. "Ursula Le Guin." In *Alternative Worlds in Fantasy Fiction*, edited by Millicent Lenz and Peter Hunt. Continuum, 2001.
Lindow, Sandra J. "Becoming Dragon: The Transcendence of the Damaged Child in the Fiction of Ursula K. Le Guin." *Extrapolation* 44, no. 1 (Spring 2003): 32–44.
McGillis, Roderick, ed. *Children's Literature and the Fin de Siècle*. Praeger (Greenwood), 2003.
Nadir, Christine. "Utopian Studies, Environmental Literature, and the Legacy of an Idea: Educating Desire in Miguel Abensour and Ursula K. Le Guin." *Utopian Studies* 21, no. 1 (2010): 24–56.
Ostry, Elaine. "'Is He Still Human? Are You?': Young Adult Science Fiction in the Posthuman Age." *The Lion and the Unicorn* 28, no. 2 (April 2004): 222–46.
Plumwood, Val. *Feminism and the Mastery of Nature*. Feminism for Today series. Edited by Teresa Brennan. Routledge, 1993.
Pullman, Philip. *The Amber Spyglass*. Del Rey, 2001.

Pullman, Philip. *His Dark Materials Omnibus*. Knopf, 2007.
Pullman, Philip. *The Subtle Knife*. Edited by His Dark Materials Omnibus. Del Rey, 2007.
Pullman, Philip. *The Golden Compass*. Edited by His Dark Materials Omnibus. Knopf, 2007.
Quirk, Anne. "God Knows, Philip Pullman." *Horn Book Magazine*, May 1, 2008: 362–63.
Rashley, Lisa Hammond. "Revisioning Gender: Inventing Women in Ursula K. Le Guin's Nonfiction." *Biography: An Interdisciplinary Quarterly* 30, no. 1 (Winter 2007): 22–47.
Rochelle, Warren G. "The Emersonian Choice: Connections between Dragons and Humans in Le Guin's Earthsea Cycle." *Extrapolation* 47, no. 3 (Winter 2006): 417–26.
Sambell, Kay. "Carnivalizing the Future: A New Approach to Theorizing Childhood and Adulthood in Science Fiction for Young Readers." *The Lion and the Unicorn* 28, no. 2 (April 2004): 247–67.
Shohet, Lauren. "Reading Dark Materials." In *His Dark Materials Illuminated: Critical Essays on Philip Pullman's Trilogy*, edited by Millicent Lenz. Wayne State University Press, 2005.
Slusser, George, Gary Westfahl, and Eric S. Rabkin, eds. *Immortal Engines: Life Extension and Immortality in Science Fiction and Fantasy*. University of Georgia Press, 1996.
Suvin, Darko. "On U. K. Le Guin's 'Second Earthsea Trilogy' and Its Cognitions: A Commentary." *Extrapolation* 47 (2006): 488–504.
Vaclavik, Kiera. "The Truth in Lyra's Lies." In *The Golden Compass and Philosophy: God Bites the Dust*, edited by Richard Greene and Rachel Robison. Open Court, 2009.
Warren, Karen J. "The Power and the Promise of Ecological Feminism." *Environmental Ethics* 12, no. 2 (Summer 1990): 125–46.
White, Lynn Jr. "The Historical Roots of Our Ecologic Crisis." In *The Ecocriticism Reader*, edited by Cheryll Glotfelty and Harold Fromm. University of Georgia Press, 1996.
Wilson, E. O. *Biophilia*. Harvard University Press, 1984.
Wood, Naomi. "(Em)bracing Icy Mothers." In *Wild Things: Children's Culture and Ecocriticism*, edited by Sidney I. Dobrin and Kenneth B. Kidd. Wayne State University Press, 2004.
Zipes, Jack. *Relentless Progress: The Reconfiguration of Children's Literature, Fairy Tales, and Storytelling*. Routledge, 2009.

Conclusion

As is probably the case for most people who have spent years in myopic proximity to a single idea, I sometimes suspect myself of having made it all up. Are there really such things as biophilic impulses and proclivities? Do the conditions of our evolution as a species really guide or perhaps even determine our actions, perceptions, and desires? Even if so, how would a particular literary genre do all the things I claim for children's literature—harbor an evolutionarily directed worldview, foster it, transmit it, express it?

Like the fish that theorized about water, at intervals I look around and remember that the thing I'm talking about is invisible and that its existence can be argued for only inductively, from observing the multiple effects of the theorized cause.

This is a way of saying about my thesis, as in my Introduction I cited Alexander Rosenberg as saying about sociobiology at large, that it "suffers not from falsity, but from obviousness."[1] Of *course* children's books are full of animals and depictions of the outdoors: what else would they be full of? Of *course* child readers are going to be attracted to books featuring children interacting with other animals as well as with their own kin: why wouldn't they be? These things are as obvious as the fact that rich Manhattanites will compete for penthouse suites with views of Central Park but not for windowless basement suites.

It is only when I start applying what I consider the plausible biological and sociobiological explanations—in the child readers' case, that they are reassured by the depicted compatibility of human ultrasociability and integration with nature; in the Manhattanites' case, that we are a bipedal, vision-oriented, generally predatory species whose ideal habitat will duplicate as far as possible the African savannah—that I mentally run up against the assumed skepticism of my Humanist/Humanities-oriented reader. At those times, I come close to wondering if I am hanging clothes on an imaginary emperor.

DOI: 10.4324/9781003784210-6

Glen A. Love, arguing for literary critics and scholars to adopt a more "inhumanist" mode of critique—one that (in the words of the term's coiner, Robinson Jeffers) rejects "human solipsism" and recognizes "transhuman significance"[2]—acknowledges that academe continues to make this a "repressed counterpoint" to humanism. Still, he holds out hope that biologically grounded viewpoints will eventually prevail, even in the corridors of the Humanities, as the "steadily increasing body of evidence" gives them "greater explanatory capability" than other theories.[3]

It is this powerful "explanatory capability" that first led me to my argument and sustains my faith in it. Given the premise that humans evolved as animals within nature and that the rupture from the non–human-made environment is unfamiliar, distressing, and in evolutionary terms very recent, it makes great sense to me that writers would seize upon the "permission-giving" granted by the acceptance in children's literature of the Romantic child, and the "cover" granted by the genre's marginalized status, to simultaneously express, transmit, and conceal their own deep passions about the natural world and our place in it.

Likewise, the notion that children have an inborn, hardwired fascination with other animals and an innate desire to interact with the outdoors (proclivities shared by adults—but with the crucial difference that children are allowed and expected to show it openly) to my mind does a much better job of explaining the popularity of children's books about animals and nature, and the resulting superabundance of these in children's literature, than do, say, Freudian theories of child development that rely on invention rather than induction.

Looking at children's literature as a "carrier" of biophilia, and as holding within it a discrete culture of syncretic biotheism that reforms the idea of the afterlife and strongly rejects the idea of the "never-death," also makes sense of the presence and persistence of certain themes. I began this book with Frances Hodgson Burnett's eloquent implied critique of Cartesian dualism and Christian and secular anthropocentrism, along with her proposal that we are pantheistically and atomistically integrated into nature in ways both literal and mystical, and that through rejoining the natural cycle of the earth we rejoin each other also. In my final chapter I arrived, more than 100 years of children's literature later, with two books that, in Graham Holderness's words, "[are] imbued with a romantic pantheism in which atoms disperse from a living body at the point of death…[and] are envisaged as becoming part of nature…but also gravitating towards one another."[4]

This is a through-line that non-biologically based theories would struggle to explain. Even if we grant the conventional explanation that it stems from children's literature authors obdurately insisting on perpetuating the concept of the "Romantic child," we then need to explain why this particular concept has survived and thrived for 250 years. Further, if we are to be fair, we should consider that it is time to promote the idea of authorial insistence to that of biological persistence. This line of thinking gets us by a more tortuous route to the same root explanation as the sociobiological one of biophilia. If children's literature, like utopian literature, can be said to be about the education of desire, then the remarkable continuity between Golden Age and "post-apocalyptic" literature affirms E. O. Wilson's contention that some desires are more readily educated than others—that, as he says, we "submit to a few biases easily while automatically avoiding others,"[5] "gravitate to about the levels of complexity" found in a natural ecosystem,[6] and "learn particular things about certain kinds of plants and animals" with "quickness and decisiveness."[7]

Recognizing the strongly biotheistic bent of children's literature also holds some explanatory power outside the genre. In *Making Nature Sacred*, John Gatta's extended examination of biotheism and Romanticism in American adult literature, Gatta remarks,

> Incredibly, belief in the more-than-material "strangeness" of material nature has somehow managed to survive this country's [America's] transformation by industrialization, the demise of non-commercial agrarianism, the intellectual triumph of empirical science, and the latter-day emergence of a postindustrial consumer society wedded to technology.[8]

That a sense of "numinousness" and reverence has survived in American adult literature, and arguably even more so in Western adult literature elsewhere, would indeed be incredible if we conceived of cultural transmission as a freestanding human social process, isolated from and unaided by evolutionary forces, rather than as a component of human biology that gives certain cultural proclivities a strength and endurance beyond their apparent powers. And even with these deeper biological proclivities acknowledged, it would remain difficult to credit if we did not note also that, for instance, Laura Ingalls Wilder's strongly biotheistic Little House series has sold around 60 million copies;[9] E. B. White's subtly but deeply reverent *Charlotte's Web* approximately 45 million;[10] J. R. R. Tolkien's pantheistic and animistic *Lord of the Rings* a reported 150 million;[11] and so on. To my mind, if we are looking for mechanisms of the "somehow," these are the most likely ones. In other words, surely we can better explain the presence and persistence of nature-worship in orthodox Western adult literature in light of its ubiquity in Western children's literature, and this ubiquity in turn surely can be plausibly explained only by evolutionary and biological theories of culture.

If E.O. Wilson is right, if evolutionists are right, and if I am right in hitching my wagon to their star, then things will continue in children's literature regardless of trend. Nature will continue as a subject and theme. Biophilia will continue to make itself felt on the page and in our lives. The syncretic biotheism that arises from it will organically and organismically find a way to persist.

On the face of it, none of this seems likely. As Jack Zipes saliently notes, "The best-selling books for young readers between three and eighteen are now produced and marketed globally in a manner that has altered the reading and reception of children's literature,"[12] and "the cultural field of production has increasingly become influenced by market conditions to such an extent that the child's capacity to read without interference from the media and advertising has become next to impossible."[13] Among other implications, this new reality means that even within the span of time in which I wrote this book, many of the backlist books and authors that were readily available on bookstore and library shelves at the start of it—Le Guin's various series, the Little House series, Cynthia Voigt's Dicey books, Diana Wynne Jones, Natalie Babbitt—have by now been bumped by the proliferation of series with movie tie-ins, from children's fantasies such as Rick Riordan's Immortals series to those for older children and young adults, such as Veronica Roth's Divergent series, James Dashner's Maze Runner series, and Pittacus Lore's Lorien Legacy series. How can biotheism in children's literature survive, or biophilia continue to be fostered and transmitted through this means, if parents and children can't even find a copy of *Charlotte's Web*?

Zipes remarks, as well, on implications for fantasy in an era when the most radical literary imaginings are "outstripped by real existing conditions" and "cannot keep pace with the devastating and disturbing fantastic of real occurrences."[14] He believes that, in reaction, books and films for children and young adults have taken on an essentially placatory role: "[W]e turn to...all kinds of fantastical artworks for diversion, what the French call divertissement, to take our minds of reality."[15] Ursula K. Le Guin, similarly, perceived that people turn to fantasy works when they are afraid of change, seeking reassurance in it "stability, ancient truths, immutable simplicities ...[which] the mills of capitalism provide." She believed that "commodified fantasy"—which children's and young-adult fantasy certainly has become in the past decade or so—"invents nothing, but imitates and trivializes."[16] It is possible that the age of hyper-accelerating technology and consumerism, abetted by the pharmacological power to numb rebellious primal-stem impulses, will overwhelm the ability of any medium or genre to express our more animalistic impulses and affiliations.

That said, I see signs that these impulses and affiliations are even now finding their way through the obstacles of commodification and to young readers. "The Harry Potter" books, a publishing phenomenon that will not soon be relegated to the scrapheap of history, reaffirmed a Tolkienesque pastoral and animism as commercially successful, at the same time that film adaptations of Tolkien's books themselves were making that author a publishing phenomenon all over again. There is a marked resemblance between James Fenimore Cooper's Natty Bumppo/Hawkeye and The Hunger Games' Katniss Everdeen, with her intimacy with nature and survivalist skills (tracking, hunting, path-reading). And while the source of the Hunger Games' phenomenal success may have been widely misidentified by publishers, parents, critics, and the readers themselves, and the evolution-based appeal left out of the knockoffs mentioned above (Roth, Dashner, Lore and others), there is nonetheless a distinct technophobia even in these more ecologically impoverished books that we could see as the flip side of biophilia.

The force of biophilia is enough, at any rate, to make the humanist and posthumanist contingent uneasy. Pro-technology scholar Noga Applebaum laments that

> fiction written for children...engages with technology... with the intention of "exorcising" modern manifestations of the machine. Perceptions of technology as a corrupting force, particularly in relation to young people's use of it, are related to the prevailing myth of the innocent child...and result in fiction written for a young audience which endorses a technophobia agenda.[17]

As more children's and young-adult books move to themes of posthumanism, it seems that Applebaum is right. Contrary to Kay Sambell's claim that "children's dystopian texts" position their "warnings about destructive tendencies in human behavior...most specifically in its [a]nimal characteristics...predation, greed, rapacious short-term opportunism, manic and irrational competitiveness,"[18] or Elaine Ostry's that in posthuman books for young people "[c]hoosing to be human means rising above matter, to put one's foot rather firmly in the 'nurture' camp of the nature/nurture debate [and] getting further away from biology,"[19] the "technophobic" take on posthumanism in even the most nature-free of these books suggests that it is exactly the loss of our animal, evolutionary, biological identities against which they are fighting. Thus, even the latest trends in publishing for young people seem to be part of a century-long tradition of representing, promoting, and (increasingly) defending our identity as animals with rightful places in both evolution and the natural world.

In my Introduction, I outline the possible reasons that children's literature and ecocriticism have not yet struck up the partnership between material and critical mode that seems so inevitable and so self-evidently right.

Although I propose the main cause to be the conflict between children's literature critics' strongly theoretical bent and ecocriticism's more reality-engaging origins, as my project has gone on, I have become increasingly convinced that the two really only need to be turned toward one another in order for the partnership to come about of its own accord. Perhaps, in illuminating the sociobiological and biospiritual nature in a representative range of children's fantasies, this book will help bring the genre and the critical framework into each other's visible spectrum. Perhaps it will equally encourage ecocritics to begin delving into the immense trove of children's literature and children's literature scholars to consider approaches of ecological literacy and evolutionary psychology. If, as I expect to happen, the floodgates of children's literature ecocriticism open in the near future, I can only hope that the academy treats the genre's biospirituality respectfully and, in the event that exposure seems to be harming this vital source of reverence, is willing to turn away and release it back into obscurity.

Notes

1 Alexander Rosenberg, *Sociobiology and the Preemption of Social Science* (Johns Hopkins University Press, 1980), 155.
2 Glen Love, *Practical Ecocriticism: Literature, Biology, and the Environment* (University of Virginia Press, 2003), 23.
3 Love, *Practical Ecocriticism*, 8.
4 Graham Holderness, "'The Undiscovered Country': Philip Pullman and the 'Land of the Dead,'" *Literature and Theology* 21, no. 3 (2007): 287.
5 E. O. Wilson, *Biophilia* (Harvard University Press, 1984), 101.
6 Wilson, *Biophilia*, 79.
7 Wilson, *Biophilia*, 85.
8 John Gatta, *Making Nature Sacred: Literature, Religion, and Environment in America from the Puritans to the Present* (Oxford University Press, 2004), 4.
9 Anita Clair Fellman, *Little House, Long Shadow: Laura Ingalls Wilder's Impact on American Culture* (University of Missouri Press, 2008), 5.
10 Harvey Freedenberg, "The Story of Charlotte's Web," *Harrisburg Magazine*, May 8, 2013. https://harrisburgmagazine.com/the-story-of-charlottes-web/#:~:text=With%20sales%20of%20more%20than,of%20both%20youngsters%20and%20adults.
11 "Timeline 1954," *HarperCollins*, accessed March 6, 2026. https://200.hc.com/timeline/1954-the-lord-of-the-rings-by-j-r-r-tolkien-is-published/.
12 Jack Zipes, *Relentless Progress: The Reconfiguration of Children's Literature, Fairy Tales, and Storytelling* (New York and London: Routledge, 2009), 2.
13 Zipes, *Relentless Progress*, 4.
14 Zipes, *Relentless Progres*, 46.
15 Zipes, *Relentless Progres*, 48
16 Ursula Le Guin, *Tales of Earthsea* (Ace, 2003), xiv.
17 Noga Applebaum, *Representations of Technology in Science Fiction for Young People* (Routledge, 2010), 1.

18 Kay Sambell, "Carnivalizing the Future: A New Approach to Theorizing Childhood and Adulthood in Science Fiction for Young Readers," *The Lion and the Unicorn* 28 (2004): 246.
19 Elaine Ostry, "'Is He Still Human? Are You?': Young Adult Science Fiction in the Posthuman Age," *The Lion and the Unicorn* 28, no. 2 (2004): 240.

Bibliography

Applebaum, Noga. *Representations of Technology in Science Fiction for Young People*. New York and Abingdon, UK: Routledge, 2010.

Fellman, Anita Clair. *Little House, Long Shadow: Laura Ingalls Wilder's Impact on American Culture*. Columbia, MO: University of Missouri Press, 2008.

Freedenberg, Harvey. "The Story of Charlotte's Web." *Harrisburg Magazine*, May 8, 2013. Accessed March 6, 2026. https://harrisburgmagazine.com/the-story-of-charlottes-web/.

Gatta, John. *Making Nature Sacred: Literature, Religion, and Environment in America from the Puritans to the Present*. Oxford: Oxford University Press, 2004.

Holderness, Graham. "'The Undiscovered Country': Philip Pullman and the 'Land of the Dead'." *Literature and Theology* 21, no. 3 (September 2007): 276–92. https://doi.org/10.1093/litthe/frm025.

Love, Glen A. *Practical Ecocriticism: Literature, Biology, and the Environment*. Charlottesville and London: University of Virginia Press, 2003.

Minzesheimer, Bob. "'Charlotte's Web' at 60: By the numbers." *USA Today*, November 4, 2012.

Nadir, Christine. "Utopian Studies, Environmental Literature, and the Legacy of an Idea: Educating Desire in Miguel Abensour and Ursula K. Le Guin." *Utopian Studies* 21, no. 1 (2010): 24–56.

Ostry, Elaine. "'Is He Still Human? Are You?': Young Adult Science Fiction in the Posthuman Age." *The Lion and the Unicorn* 28, no. 2 (April 2004): 222–46.

Rosenberg, Alexander. *Sociobiology and the Preemption of Social Science*. Baltimore and London: Johns Hopkins University Press, 1980.

Sambell, Kay. "Carnivalizing the Future: A New Approach to Theorizing Childhood and Adulthood in Science Fiction for Young Readers." *The Lion and the Unicorn* 28, no. 2 (April 2004): 247–67.

Wilson, E. O. *Biophilia*. Harvard University Press, 1984.

Zipes, Jack. *Relentless Progress: The Reconfiguration of Children's Literature, Fairy Tales, and Storytelling*. New York and London: Routledge, 2009.

Index

Note: Page numbers with "n" refers to footnote numbers

www.ingramcontent.com/pod-product-compliance
Ingram Content Group UK Ltd.
Pitfield, Milton Keynes, MK11 3LW, UK
UKHW021028031026
476244UK00004B/58
* 9 7 8 1 0 4 1 3 2 7 2 8 8 *